FOREVER SOMEWHEN

The Immortality of Space, Time and You

By Patrick Wyliams

This second edition is published in 2026 by
Old House Publishing,
13 Adelaide Road,
Dublin 2,
D02 P950,
Ireland

First published in 2025 by Old House Publishing.

CONTENTS

INTRODUCTION

This universe may end one day. That ending may not only be inevitable; it may be necessary, and it may even be desirable, because it could be the very thing that allows existence, including yours, to continue. That is a strange claim to open a book with, and I don't expect you to accept it yet. I only ask that you hold it loosely for a few hundred pages and see whether it still looks strange at the end.

What follows takes some of the most alarming ideas in modern science, the death of the universe chief among them, and looks at them again from a wider vantage. Endings tend to frighten us because we meet them at human scale, where an ending is a loss and nothing follows it. Seen from further out, an ending can be a hinge, not a full stop. We will follow several serious scientific theories to see where they lead, and some of them lead somewhere surprisingly consoling.

To get there we'll need to think in more dimensions than the four we live in. That word carries a lot of unhelpful baggage, most of it supplied by science fiction, so I'll say plainly at the outset that nothing in this book involves invaders arriving from another dimension. A dimension, in the sense used here, is simply a way of measuring something. We will start with the four we know, spend a good while examining the fourth, which is time, and then climb steadily outward. Along the way we'll build up the vocabulary we need for the larger structures. By the end we'll have a name for the whole of it. I call it the Omni-Verse, meaning the totality of everything that exists, including realities that may run on entirely different rules from ours. I'm not going to define the intermediate terms here, because a definition offered before you need it is just a word to memorise. They will arrive when the argument requires them.

There is one habit of mind worth establishing before we begin, and it is this: our senses were built for staying alive on a particular planet, not for perceiving reality in full. Ultraviolet light is a good example. It surrounds you constantly, it is entirely real, it burns your skin on a bright day, and you can't see a photon of it, because human eyes never evolved the equipment. Bees can see it, and the flowers they pollinate carry ultraviolet patterns that are invisible to us and unmistakable to them. Nothing about that is mysterious. It is simply a case of the world containing more than one particular animal is equipped to register.

I raise it because the same may well be true of time. We experience time as a one-way street running from a fixed past into an open future, and that experience is so total, so continuous, and so useful that it feels less like a perception than like a fact about the universe. It may be neither. It may be the temporal equivalent of not seeing ultraviolet, which is to say a limit of the instrument, not a property of the thing being measured. Much of this book follows that possibility to see where it goes.

Let me be clear about what this book isn't. It isn't a spiritual text. It neither offers nor refutes any religious ideology, and you won't find me attacking anyone's beliefs. It is also not written exclusively for people with scientific training. I use proper scientific terms throughout, because vague language produces vague thinking, but I explain each one the first time it appears and there is a glossary at the back for anything you want to check later. You don't need a degree in physics or philosophy to read on. Curiosity and a tolerance for uncertainty will get you the whole way.

Nor is this a book that hands you a single answer. It combines established science, emerging theory, philosophy and a certain amount of ordinary critical thinking, and it puts them in conversation with one another instead of arranging them into a doctrine. Some of the models ahead sit awkwardly beside each other. A few of them can't all be true at once. That is deliberate, and the first chapter explains why I think it's a strength and not a failure of nerve.

What I hope you take from it isn't agreement. It is a slightly different posture toward not knowing. We live in a universe we don't fully understand and may not be constructed to fully understand, and it was certainly not designed around our wish to make sense of it. Space and time govern every hour of your life and remain, at the level of what they actually are, open questions. Time may not be linear. Space may not be fixed. This universe may be one of a great many.

If, by the last page, you find yourself a little less afraid of endings, whether of the universe, of humanity, or of your own life, then the book has done what I wanted it to do. Our days are crowded with small anxieties and smaller quarrels, and sometimes the widest possible frame is exactly what makes the next step manageable.

So let's begin, starting where the strangeness starts, with the nature of time.

1. BUILDING AN IDEA-SYSTEM

Imagine you're adrift at night, floating in open ocean beneath the stars. Below you the water goes down further than any light reaches. Around you the currents cross and shift, running in several directions at once, so even the sea can't agree with itself about which way things should go. Then a plank of wood knocks against your shoulder and you grab it. Of course you do. When you're adrift in deep water, a plank feels like salvation.

Belief systems are built for that moment. A religion, a political ideology, a philosophy, even a favourite scientific story can steady you when the waves of uncertainty get loud. The worse the storm, the tighter we hold on. There's no shame in that.

But planks have limits. A plank keeps you on the surface, which isn't the same as taking you anywhere. You can't swim with both arms while you're clinging with both hands. I say this as someone who once held on tighter than most, to several things that didn't deserve the grip I gave them. Let me tell you why.

At five years of age I came very close to being murdered. The person who did it was someone in a position of authority, someone I had every reason to trust, and they trapped me in a burning building and left me there to die. I got out. Others that day were not so lucky. I've no intention of dwelling on the details, because this isn't that kind of book, and because the details aren't the part that matters here. What matters is what it did to the shape of my thinking.

A child who survives something like that doesn't simply go back to being a child. From that day I stopped taking things for granted. That sounds almost like a virtue when you write it down, and it feels nothing like one when you live inside it. My mind learned to look around corners. It rehearsed what could go wrong, constantly, the way a muscle held tense eventually forgets how to let go.

Alongside all that came a fear of death that stayed with me for decades. I want to be exact about what kind of fear it was, because the distinction matters. I wasn't especially afraid of dying, of the moment itself, of whatever pain or fright might come with it. I was afraid of being dead. I was afraid of the long, silent, uninterrupted afterwards where I wouldn't be anywhere at all.

That day taught me a second lesson too, and this one has proved more useful. Be careful with authority. It doesn't matter whether it wears a uniform, a collar, a title, or just the confidence of someone who has never been asked to justify themselves. Be careful with any worldview that asks for your trust before it offers you a reason.

At seven I put my parents through a fairly intense interrogation about how anyone could actually know there was an afterlife. Their answers didn't convince me. I could see I was upsetting them, so I let it go, but the questions didn't let go of me. The idea that faith was a test, and that doubting was how you failed it, struck me even then as too convenient. It also seemed oddly cruel, coming from a being who was supposed to be both all-knowing and entirely loving.

By my teens the fear had hardened into something that took up a great deal of my attention, and I started grabbing at whatever certainty was in reach. I reached for religion first. Then mysticism, and a tour of alternative religions I wouldn't describe as rigorous. Then tidy philosophies, the kind that promise you an account of everything if you'll just accept one premise at the door. In my twenties I moved on to tidy scientific answers, which felt more respectable and worked in much the same way.

I'm not embarrassed by any of it now. I clung to what helped me function, and functioning wasn't a small achievement at the time.

There was a cost, though, and I couldn't see it while I was paying it. When a new idea didn't fit what I already held, I rejected it fast, and I told myself I was being discerning. Psychologists call this confirmation bias, the habit of noticing evidence that agrees with you and finding reasons to dismiss the rest. From inside the bubble I'd built, it didn't feel like bias at all. It felt like self-protection. I stopped reading books that made me uncomfortable. I skipped past theories that rattled my worldview. My world got smaller, and I experienced that as safety.

What I kept finding, over and over, was that too much of what I believed rested on a leap I'd agreed not to examine. That's what a belief system asks of you. It wants the whole structure accepted together, because the parts hold each other up, and pulling on one loosens the rest. Question a single beam and you'll be told, sometimes gently and sometimes not, that you're questioning the building.

So let me be clear about what this book isn't. It isn't a religious text. It neither promotes nor attacks any faith. As it happens I'm an atheist, and I only mention it so you know where I'm standing. I'm not selling it, and you won't find me going after anybody's beliefs in these pages. What follows is a blend of science, philosophy and ordinary critical thinking, aimed at some very large questions, and meant to help you reach your own answers instead of adopting mine.

Here's the frame I'd like to offer you in place of a belief system. I call it an **idea-system**, and the difference isn't decorative.

A belief system needs its parts to be true together. An idea-system cares about whether an idea is useful, not whether you've sworn loyalty to it. You take an idea, try it, test it against what you know and what you can see, keep what earns its place and set aside what doesn't. It's closer to renting ideas than owning them. You get the use of the thing without the mortgage, and when it stops serving you, you hand back the keys without feeling you've betrayed anyone.

Most importantly, in an idea-system a contradiction is a tool, not a crisis. Two incompatible pictures can sit on the desk together, each useful for a different job, and neither one has to be executed for the other to live.

That sounds like a convenient dodge, so let me show you it working somewhere nobody could accuse the participants of being sloppy.

For most of the nineteenth century, physicists were confident that light was a wave. They had good reason. Shine light through two narrow parallel slits onto a screen behind, and you don't get two bright stripes, which is what you'd expect if light were a spray of tiny bullets. You get a whole series of bright and dark bands, called interference fringes. They appear because the light arriving from one slit adds to or cancels out the light arriving from the other. That's what waves do. Drop two stones into a still pond and watch where the ripples meet, and you'll see the same pattern of reinforcement and cancellation drawn on the water. The two-slit result was treated as settled, and reasonably so.

Then in 1905 Albert Einstein published an explanation of something called the photoelectric effect. Shine light on certain metals and it knocks electrons loose. The puzzle was that brightness didn't decide whether electrons came free. Colour did. Dim blue light would liberate electrons while intensely bright red light wouldn't, and that makes no sense at all if light is a wave delivering energy smoothly. It makes perfect sense if light arrives in discrete packets, later named photons, each carrying an amount of energy set by its colour. One packet with enough energy knocks one electron loose. A million packets without enough energy knock nothing loose, however dazzling the beam. That explanation eventually won Einstein a Nobel Prize, and it describes light behaving exactly like a stream of particles.

So light is a wave, and light is a particle, and those two descriptions can't both be complete.

Modern versions of the two-slit experiment sharpen the problem instead of solving it. Send light through the slits one photon at a time, so faintly that only a single packet is in the apparatus at any moment. Each photon lands on the detector as one small point of impact, which is particle behaviour of the plainest kind. Leave it running long enough and those individual points build up into interference fringes, which is wave behaviour of the plainest kind. Every photon arrives as a dot. Together, the dots draw a pattern no collection of bullets should ever draw.

The Danish physicist Niels Bohr gave this situation a name in 1927. He called it **complementarity**. The wave picture and the particle picture are both necessary, neither is sufficient on its own, and no single experiment will ever show you both at once. Physicists didn't settle it by holding a vote. They learned to reach for whichever description the job in front of them needed, and they got on with the work.

This example is easy to overstate, and it often is overstated, so let me be exact. Nature isn't confused, and quantum mechanics isn't self-contradictory. The mathematics underneath all of this is a single consistent theory that has survived a century of the most punishing tests we can devise, and it doesn't stutter between two answers. The contradiction lives in our pictures, not in the world. Waves and particles are both ideas we imported from ordinary human-scale experience, from water and from thrown stones, and light is under no obligation to resemble either. What those physicists were doing was holding two imperfect models at once, using each where it earned its keep, and declining to pretend either was the final word.

That's an idea-system, practised at the highest level of rigour our species can manage. If it's good enough for the people who built the theory behind the transistor and the laser, it's good enough for a book about time.

I bring it up now because you're going to need it. Some chapters will build carefully on a model, and then a later chapter will lean on that same model until it creaks. In Chapter 3 I'll make a case for one picture of how time might turn back on itself, and in Chapter 4 I'll set that picture beside a rival that can't be true at the same time. Later still, when we reach the fifth dimension and the branching of possibilities, we'll meet ideas that sit uncomfortably beside the fixed, finished universe of Chapter 2.

None of that is an accident, and none of it is me losing the thread. We're mapping a coastline, not paving a straight road, and a coastline has inlets that double back. Where two of my models really do conflict, I'll say so out loud instead of smoothing it over and hoping you don't notice. What I'd ask in return is that you resist picking a winner too early. The interesting work usually happens in the space between two ideas that can't both be right. Close that space off in Chapter 3 and you'll have nothing left to think with by Chapter 9.

If a chapter leaves you unconvinced, that's a perfectly good outcome. No oath is required here. The book does reach conclusions, some mine and some borrowed, and I'll weigh their implications as fairly as I can while admitting where new evidence could overturn them. There are ideas in here I don't expect to be proved true, and I hold those loosely. There are others I hold firmly that you may well think are rubbish, and you're entitled to. A flexible frame lets us go a long way into deep questions without pretending to a certainty none of us has, and it keeps the structure standing when a beam has to be replaced.

I said earlier that I was afraid of being dead. That was true for a long time, but it wasn't my only fear. One of my deepest fears was that I would give in to nihilism.

What frightened me most was the suspicion that meaning would evaporate the moment I looked at it properly, and that understanding the universe would drain it of everything that made it worth understanding. I worried in earnest that knowledge and wonder were in competition, and that the more I learned the less there'd be to feel.

For a while I even imagined becoming a priest. It would have given me a life spent among the largest questions there are. I was far too curious for it, and far too full of the wrong sort of questions, and I knew any honest week in the job would have collapsed under the weight of them. When faith went hollow, the anxiety came straight back through the gap it left. Other systems promised comfort and delivered rather less than advertised.

What eventually helped came from an unglamorous direction. I started meditating, which for me meant learning to make room for the body's alarm bells without letting them take the wheel. In the harder stretches I practised loosening my grip on the idea of a fixed and permanent self, gently, without forcing it. Then I did the thing I'd spent twenty years avoiding. I accepted that I'm going to die, and I let that fact sit beside me without arguing with it.

Something changed at that point, and it changed in a direction I hadn't predicted. Letting go of the demand for a final answer began what I've come to think of as an **Existential Detox**. I stopped needing every idea to reassure me before I'd let it in the door. I went back to books I'd abandoned and theories I'd found threatening. This time I wasn't bargaining with them. I was just looking.

Freed of the fear, I found black holes and big bangs were beautiful. Theory didn't deepen my nihilism at all. It began, slowly, to dissolve it. The contradictions of reality, once I could hold them without panicking, stopped being threats and started being patterns worth admiring. I stopped needing a single plank, and I learned instead to move between conflicting theories, keeping whichever parts proved useful and letting the rest drift.

Along the way I noticed something about clever people, and I noticed it in the religious and the scientific alike. There's a territorial reflex that comes over us when an idea becomes ours. Theories get guarded like property instead of used like tools, defended long past the point where defending them is reasonable, because an attack on the theory has quietly become an attack on the person.

I recognised the reflex easily enough, because I had it too. Seeing it in myself is what finally loosened my grip. Fear narrows a mind, which is well documented and which anyone who has ever sat an exam already knows. A mind that isn't frightened can widen instead. When I stopped demanding that the universe have a grand order, I found I could actually follow the arguments I was reading. When I stopped worrying about drowning, it turned out I could swim. And when I accepted that existence was final, I started noticing all the serious scientific arguments suggesting it might not be.

Swimming costs something, though, and I've made it sound easier than it is.

A plank asks nothing of you except that you hold on. Swimming asks for continuous effort, and it doesn't let you rest the way floating does. You have to keep revising, keep testing, keep noticing when an idea you liked has stopped earning its place, and there's no moment where the work is finished and you're allowed to stop. Swimming also means living without the particular comfort of certainty, and that's a real comfort, not a worthless one.

People who tell you it costs nothing to give up have usually not given up much.

What swimming offers is the freedom to go somewhere. It stops you spending your one life clinging to a piece of wood that was never going to take you anywhere.

That's the trade. I think it's a good one, and I wouldn't have thought so twenty years ago.

Where many science books stop at the mechanism, and many spiritual books stop at the doctrine, this one takes some major scientific theories seriously and then asks what they might mean for someone who has to get up on Monday. It isn't trying to convert you. It's trying to wake up your curiosity and offer you a different posture toward not knowing.

We live in a universe we don't fully understand and may not be built to fully understand, and it certainly wasn't designed around our wish to make sense of it. Space and time govern every hour of our lives and remain, at the level of what they actually are, open questions. Time may not be linear. Space may not be fixed. This universe may be one of a great many.

You'll meet scientific terms as we go. Sometimes I'll explain them where they stand, and the rest are waiting in the glossary at the back. Please don't try to memorise the vocabulary. There's no exam at the end of this.

If a section feels hard, swim past it and come back later, or don't come back at all. I'm after your engaged attention, not your agreement.

I've no wish to be revered, and the thought of anybody treating this book as scripture makes me want to lie down. Most of the theories ahead aren't mine. I've picked them up, turned them over, and asked what they'd mean if we took them seriously, and also what it would mean if we didn't. My ambition is to be a decent guide. I'll point out the landmarks, admit where the fog is, and remind you to watch your footing.

Ahead of us are models that bend the intuition badly. There are arrows of time that point different ways and refuse to agree. There are universes that bounce, and universes that mirror. There are selves that turn out to be more pattern than substance. Some of these ideas will comfort you and some will prise your fingers off a rail you didn't know you were holding, and both of those are worth having.

So let's begin. Not by grabbing the nearest plank, but by going over the side into deep water. We start with time.

2. SOMEWHERE, SOMEWHEN

We experience time as a one-way street running from the past into the future. Our memories sit behind us and are fixed, and what lies ahead is unknown. That arrangement isn't a philosophical position we've adopted; it is the shape of ordinary life. We plant seeds in spring because we expect to eat in autumn. We save because we expect to grow old. We apologise for what we did last week and not for what we'll do next week. It is vital that we perceive reality this way, for the simple reason that this way works.

But what if that is a fact about us and not about time? What if we experience one direction of travel because living creatures need to experience one direction of travel in order to function, and the underlying reality is less tidy? This isn't a new idea, and I'm not the person who thought of it. It is, though, an idea worth taking seriously, and the whole of this chapter is an attempt to do that.

Before we go anywhere near the fourth dimension, we should settle what a dimension is, because science fiction has done real damage here. When most people hear the word they picture something to be invaded from, a parallel place with a portal and probably a monster. The actual meaning is far more modest. A dimension is a way of measuring something. A tailor asks for your dimensions and means your measurements. A physicist asks the same question with more axes and a worse suit. When we say that reality has a certain number of dimensions, we're saying that there are that many independent ways of measuring it.

You already navigate three of them without thinking. Watch a film on a flat screen and the image has length and height, two measurements, but no depth, because it is a flat projection onto a flat surface. Put on a pair of 3D glasses and your brain is tricked into inferring a third measurement, and the picture appears to have depth. Step outside the cinema and everything actually does. Walk around a person and, like any competent tailor, you can measure their height from head to toe, their width from shoulder to shoulder, and their depth from chest to back. Do ask permission first.

We perceive three spatial dimensions because we're three-dimensional beings, which means we can move freely in all three. Anyone old enough to remember the original Pacman will recall the limits of a world with one fewer. You could move your character up or down the screen, which is height, and left or right, which is length, but you couldn't move the character toward you, out of the screen, to dodge a monster coming down a corridor. There was no such direction available. Modern games let you sidestep and duck because they add depth, and that third measurement changes the entire logic of what is possible.

Mathematics, sometimes called the universal language, holds that these three dimensions exist and are real. Nobody disputes that, because each of us can verify it in the space of about four seconds. Our direct experience confirms it, which is exactly what makes it so easy to accept. However, mathematics also holds that there are more dimensions in total, ones we find considerably harder to picture and certainly can't walk through.

How many more depends on which theory you ask. Superstring theory, which models the fundamental constituents of matter not as points but as unimaginably small vibrating strings, requires ten dimensions to work: nine of space and one of time. M-theory, which unifies the several competing versions of string theory into a single framework, adds one further spatial dimension and arrives at eleven. Neither theory has been confirmed by experiment, and both remain contested, but both are mathematically serious and widely worked on.

I want to flag something here, because it will matter later and I would rather be straight with you now than be caught out in Chapter 6. The extra dimensions in string theory and M-theory are generally thought to be compactified. That means curled up so small we never encounter them. Picture a garden hose from a long way off. It looks like a line, a single dimension. Get close, or be an ant walking on it, and there's a whole second direction available that was there the whole time. Those aren't the same thing as the ladder of alternative realities we'll start climbing later in this book. When I use dimensional language to compare possible universes, I'm borrowing a useful way of organising ideas rather than describing the compactified dimensions of M-theory, and I'll say so plainly again when we get there. Holding those two senses apart will save you a great deal of confusion.

The first four dimensions are the ones we experience without effort: length, width, height and time. Time is the fourth. You existed a minute ago and you exist now, and we measure ourselves in it constantly, saying we're a day older. You are the same person as yesterday, and you're also a day further along. If we didn't move through time, we would exist for a single instant and then blink out, and conversation would become difficult, to put it mildly.

For now we're stopping at the fourth and having a proper look around. Time is something we can measure closely and travel through continuously, but apparently only in one direction. We can't walk into yesterday. We are born, we grow, we age, and we die, and the whole business feels like a description of how the universe must work.

It may not work that way at all. Remember the ultraviolet light from the introduction, real and everywhere and permanently invisible to us. The true nature of time may be one of those things we can't see clearly from inside it, in the way that a fish is poorly placed to develop a theory of water.

Consider what physics actually says. We experience a forward arrow of time, and yet many of the core equations describing motion don't prefer forward over backward. Take an equation describing how objects move under Newton's laws, replace time, written as t, with negative time, written as minus t, and the equation still works perfectly well. This is called time-reversal symmetry. It doesn't hold everywhere, and there are known and important exceptions involving certain rare particle interactions, but it is far more common than our experience would lead you to expect. The underlying laws, in other words, don't much mind which way time points, even though we very much do.

Before we start questioning the direction of time, though, it's worth asking about its nature, because that turns out to be where the existential weight sits.

Very few people are troubled by their non-existence before birth. Almost everyone is troubled by their non-existence after death. That asymmetry is worth sitting with, because the two states are, on the face of it, identical. Some people respond by holding tightly to a rigid belief system and defending every part of it, sometimes fiercely, because the alternative feels like an abyss. Others adopt a determinedly bleak position, insisting that nothing matters and that science offers only a cold ending. In both cases the fear is doing the steering, not the thinking.

A great deal rests on one assumption, which is that time has a single preferred direction and that only the present moment is real. If that is true, then when "now" arrives at your death, that really is that. But physics doesn't require the assumption. It may simply be our one-way perception shaping our sense of what exists.

Here is a way to feel the asymmetry. You aren't distressed that you aren't currently standing on the Moon, because you're here instead, and here is a real place. Nor were you distressed, before you were born, about not being born, because there was no "now" positioned there to do the distressing. This points toward a larger idea, one supported by some interpretations of relativity and by several serious philosophical models of time:

All times exist.

Stated more plainly: the past still exists, and the future already exists. All of us are travelling through the fourth dimension, and we experience one moment at a time, but the moments we aren't experiencing haven't thereby stopped being. Why should our experience be so restricted? Most likely because life requires that kind of stability. A universe in which creatures could wander the timeline at will would be chaotic beyond function, and the very beings capable of such time tourism would probably never evolve in the first place. So we get a useful constraint, which is a one-way subjective experience that keeps life coherent.

Our everyday sense that only the present is real has a name in philosophy. It is called presentism. It is entirely natural, and it doesn't sit comfortably with what relativity tells us about spacetime.

It can be very hard to accept that what we perceive may not reflect what is actually there. Try telling someone committed to a flat Earth that the planet is round, and you'll see how reluctant people can be to accept a truth that contradicts their own eyes. The rest of us aren't immune; we simply happen to be right about that particular question.

There are two main alternatives to presentism, and they differ in one important respect.

The first is Growing Block Theory. On this view the past and the present both exist right now, but the future doesn't exist yet. Reality is like a block that is still being built, with each new moment added to the top of the pile while what lies ahead remains undetermined until it arrives. It has an intuitive appeal, though it does carry the odd implication that this moment, the one you're reading in, is the moment of creation, and that the same was true of every moment before it.

The second is Block Theory, also called Eternalism, which proposes that past, present and future all exist equally and already. Time is a dimension, not a process. Just as every location in space exists whether or not you happen to be standing in it, every point in time exists whether or not you happen to be experiencing it.

I mentioned that we don't fear the moments before our birth. That is because we exist now, and because we believe those earlier moments won't come round again. We face along the arrow of time and we fear what is still to come. But if the past still exists and tomorrow already exists, even though you can only ever experience "now", then you exist in every moment you ever occupy.

It is difficult to think clearly about something you've no word for. It is said that the Inuit can recognise many different kinds of snow, in part because they have many different words for them, where to me it is all simply snow. That is the nature of language. We use it to describe what we see, and it also shapes what we're able to see, because the habit of naming a distinction trains the mind to keep noticing it.

With that in mind, I would like to coin a term. I use **Somewhen** to refer to a moment in time considered as a location, a place where things exist, rather than as an instant that has been used up and discarded. Somewhen else, you're four years old. Somewhen else, you may already be a hundred.

Take a concrete case. One-year-old you is alive and well, surrounded by the people who loved you then. Five years from now, one-year-old you'll still be alive, in the moments belonging to that slice of time. Five million years from now, exactly the same is true. By that reasoning, the people you've lost are still alive somewhen, at a different location along the fourth dimension.

If time is a landscape, why do we only ever face one way?

Step outside the story for a moment and imagine the entire universe as a reel of film, with its beginning and its ending already printed on the strip. You are one of the characters on it.

Thread that reel into a projector and press play, and you experience your life exactly as you remember experiencing it. Moments arrive one after another, always forward. You have no access to the whole reel. You only ever see the frame you're in, plus your memory of the frames that came before.

Press pause on any single frame and you still exist there. Your life is spread across a vast number of frames, and your awareness occupies one of them at a time.

Now let me do something mischievous and press rewind. On the screen, ink lifts off your signature and gathers back into the pen. Footprints leap out of wet sand. A spilled glass hauls itself upright and the water climbs back in. At your fifth birthday, torn paper rises from the floor, stitches itself along its seams and wraps the box again. The toy slides out of your hands and back into its packaging. You hand the gift away. The grin on your face fades back into curiosity, and then into the serious concentration of a child pretending very hard not to know what is coming.

Here is the part that matters. Inside each frame, you still feel time moving forward. Consciousness has no reverse gear. Even with the projector running backward, the you in any given frame remembers only earlier frames and anticipates only later ones. Your subjective arrow points forward in every slice of the film, whichever way I turn the reel.

That is one way to picture the claim that all times exist. The reel is already there, past scenes and future scenes together on the strip. We don't see the whole of it at once because creatures like us appear to need a one-directional experience in order to function at all. The scenes themselves don't wink out of being when we leave them. They stay on the reel, and each of them is a somewhen you occupy.

Now for the part that is unsettling and consoling in roughly equal measure. If time has this block-like structure, then you're already dead somewhen, both before and after your lifetime. You are also permanently alive in every moment of that lifetime. The longer you live, the more moments you get to keep.

Does that mean grief has no place? It doesn't. In nineteenth-century Ireland, families held wakes for relatives emigrating to America, and those wakes weren't a confusion about who was dying. The families knew perfectly well that the person was still alive and would go on living. They were mourning the end of shared presence, of the ordinary overlap of two lives in the same place at the same time. That's what we mourn now. Grief marks the end of overlapping somewhens, and it honours something real.

There is comfort available alongside it. Living well means those good moments are lived again and again, each time as if for the first, because from the inside there's no other way to live them. In a single-universe view you would always make the same choices in those moments, and I don't think that makes them meaningless. I think it makes them permanent. What you choose now isn't swept away behind you. It is kept.

Think of time as an archivist, not a reaper, gathering each moment and filing it carefully. The last embrace you shared with someone you love is a permanent page in the archive. The first laugh you shared with them is happening somewhen else. Somewhen else you're tasting ice cream for the first time, again. It isn't often that a scientific idea offers this sort of consolation, and I think we should take it when it comes.

There is also harder physics pointing the same way. In Einstein's theory of relativity, simultaneity is relative, which means that two observers moving differently can disagree about whether two events happened at the same time, and neither of them is mistaken.

Picture a long train rushing past a station platform on a dark night. Two bolts of lightning strike the track, one at the front of the train and one at the rear. An observer standing on the platform, exactly halfway between the two strikes, sees both flashes arrive at the same instant and reasonably concludes that the strikes were simultaneous.

Now put a second observer in the exact middle of the moving train. From the platform's point of view, that observer is travelling toward the front strike and away from the rear one, so the light from the front has slightly less ground to cover and the light from the rear slightly more. The observer on the train sees the front flash first and the rear flash second, and reasonably concludes that the strikes weren't simultaneous.

Which of them is right? Both of them are. In relativity, simultaneity depends on your state of motion. Different states of motion slice up spacetime into different definitions of "now", and there's no single universal present that everyone is obliged to agree on.

If that sounds like a curiosity with no practical consequences, consider that you rely on it daily. The satellites in the global positioning system run at slightly different rates from clocks on the ground, because both their motion and the weaker gravity at their altitude affect the passage of time. Their signals are corrected using relativity, and without those corrections your phone's estimate of your location would drift badly within a day. Seen in that light, talking about somewhens, treating times as places, stops being a poetic flourish and becomes reasonably practical language for how spacetime behaves.

If there's no single universal now, then it makes rather less sense to say that the past is gone or that the future isn't yet real. All of time forms a four-dimensional spacetime block.

What does any of this mean for daily life? You still have to put the bin out and go to work. What it changes is your sense of what you're. Your life isn't only unfolding; it is also written across time. You aren't merely a moment in motion. You are a being extended through time, even though you can only experience it one slice at a time. Just as a character exists in every frame of a film, you exist in every somewhen you occupy.

Somewhen is an idea I've folded into my own idea-system, mostly because it helps in loss. It doesn't erase grief, and I would distrust anything that claimed to. What it does is remind me that somewhen I'm walking a winding country road with my grandmother, nine years old and confidently debating the meaning of life, and she is laughing back at me. Yes, I was that sort of child. A somewhen shared lasts.

None of this makes me any less interested in living a long and healthy life. People often describe ageing as losing marbles from a jar, one for each year, the jar emptying as you go. Having very nearly died at five, I prefer the image the other way round. Each birthday adds a marble. Each year adds more somewhens to the jar. If time is permanent, then this life is permanent in the sense that matters most, and I would like to fill it.

If you're wondering what any of this does to free will, hold that thought. We are currently looking at the question from a four-dimensional vantage point, which is the least forgiving one available. Later, when we reach the fifth dimension and the branching of possible outcomes, choice comes back into the picture in a way that may surprise you. For now, sit with the possibility that every moment of your life may exist permanently, including the excruciating ones. Yes, that includes the one you just thought of.

3. DISCOVERING THE REVERSE-VERSE

In Chapter 1, I promised you an idea-system, not a belief system. I'm not handing out certainties. I'm handing you interesting tools and saying, try this and see what it opens. I'm the tour guide rather than the painter, and I still occasionally lean on the rope and set off an alarm. In that spirit, let's play with a large and beautiful tension, which is the possibility that time works in two directions at once.

This idea can be built on top of Eternalism from the last chapter, or you may feel it sits awkwardly beside it. Either reaction is fine, and I would rather you noticed the friction than smoothed it over. The point of the exercise is to question and consider without being steered by fear. That is enough preamble, so let's get moving.

We begin with the Mirror Verse. Anyone who knows their Star Trek will hear "mirror universe" and picture an alternative reality where the good people are bad, the bad people are good, and Captain Kirk has acquired a goatee. Entertaining as that is, it has nothing whatever to do with the Mirror Universe Hypothesis we're exploring here.

We are staying inside the same four dimensions, three of space and one of time. We aren't changing events, and there's no evil twin. We are asking a narrower and stranger question: could there exist a universe whose overall arrow of time points the other way, while still feeling perfectly ordinary to anyone living inside it?

Call it the Reverse-Verse.

Physics has a long-standing love of symmetry, and the love isn't sentimental. As far back as Newton, we understood that for every action there is an equal and opposite reaction, and that principle is among the best supported and most thoroughly tested in all of science. Since then we've kept finding the same pattern. Flip things over, mirror left and right, run the film backward, swap particles for their antiparticles, and a remarkable number of the core laws carry on working exactly as before.

The Mirror Universe Hypothesis proposes that for every forward-moving universe there exists an equal and opposite time-reversed counterpart, which is what I'm calling a Reverse-Verse. The idea is rooted in a particular symmetry called Charge-Parity-Time symmetry, usually shortened to CPT symmetry, and this one really is a fundamental principle of quantum field theory. According to CPT symmetry, if you take a system and simultaneously invert the electric charge of everything in it, which is the C, mirror its spatial coordinates, which is the P, and reverse the direction of time, which is the T, the physical laws come out unchanged. Translated into plainer language, everything in a Reverse-

Verse would work the way ours does, only running the other way.

We can't step outside our universe and check. What we can do is look at the mathematics, and there the picture is suggestive. Conformal Cyclic Cosmology, proposed by Roger Penrose, indicates that the physics governing a contracting universe could be indistinguishable from that of an expanding one. That lends weight to the idea that time might run in both directions, with a mirrored universe evolving counter to ours while appearing entirely natural to anyone inside it.

That last point is worth pausing on, because it is the one people find hardest. A person living in the Reverse-Verse would experience time exactly as you do. Think back to the film reel from the last chapter, where the projector could be run either way and the you inside any given frame still remembered only earlier frames and still anticipated only later ones. Consciousness has no reverse gear, and neither, apparently, does memory. To them, their past is behind them and their future is ahead of them, and it is our universe that would look peculiar.

I should say plainly that I find this theory persuasive, and I think it may help explain several anomalies we observe in the visible universe. We will come back to that. First we need to establish whether the idea is compatible with known physics, how it interacts with entropy and gravity, and only then how two mirrored timelines might relate to one another.

In the last chapter we met the arrow of time, our sense that time moves forward, and asked why it appears to run in only one direction. The laws of physics seem largely indifferent. And yet in ordinary life we never see time reverse. Ice melts and never unmelts. Because we only ever see it happen one way, we infer that this is the preferred state of the universe.

We touched on this before, but a gentler recap is worth having, because physicists actually distinguish between two arrows.

The psychological arrow of time is how we experience time, with memories behind us and uncertainty ahead of us. The thermodynamic arrow of time is how entropy behaves. In an isolated system, entropy tends not to decrease. Perfume spreads through a room. A dropped cup shatters. Ice melts on a warm afternoon. You never once watch the shards hop back together and the cup leap into your hand.

The link between the two arrows is intimate. We remember the past and not the future because the past is the direction in which records exist: footprints in sand, photographs on phones, neural traces in brains. Every one of those records is a small pocket where entropy increased in the act of writing it. A memory is a debt paid in disorder. So the behaviour of entropy is a large part of why our inner arrow points the way it does, and it deserves a closer look.

In the first tiny fraction of a second after the Big Bang, far less than a billionth of a billionth of a billionth of a second, the universe was extraordinarily hot, extraordinarily dense and very nearly uniform. During that sliver of time it inflated at an exponential rate. Despite that initial evenness, matter didn't go on to condense into one tidy orderly mass. It distributed itself unevenly, forming regions of concentrated matter separated by vast stretches of near-emptiness.

The non-uniformity is thought to have arisen from quantum fluctuations, which are tiny variations in energy occurring at the smallest scales, and which were stretched across the whole universe as it expanded. Those fluctuations produced slight differences in gravitational pull, and those differences acted as seeds for structure, letting some regions grow denser while others thinned out.

Here we hit something counterintuitive, and I want to slow down for it, because the standard account skates past it and leaves readers quietly confused.

Entropy is a measure of how many different ways the parts of a system can be arranged while still looking the same from the outside. Spread perfume evenly through a room and there are a colossal number of ways to shuffle those molecules around without you noticing any difference, which is why the even spread is the high-entropy state. That is the intuition most of us carry, and it says that uniform means high entropy.

Gravity turns that intuition inside out. A large cloud of gas left to itself under gravity doesn't stay smooth. It clumps, and the clumping raises entropy instead of lowering it, because collapsing matter converts gravitational energy into heat and radiation and pours it into the surroundings, opening up vastly more ways for the total system to be arranged. A star is a tidy-looking object that is furiously manufacturing disorder. So when we say the early universe was in a low-entropy state because it was so smooth, we're speaking about a universe in which gravity is the dominant player on the largest scales. Smoothness is orderly when gravity is in the room, and messy when it isn't. Once you've that, the rest of the story stops feeling like a contradiction.

The formation of galaxies, stars and empty voids over the following hundreds of thousands of years was therefore not a defiance of entropy but an expression of it. The universe produced intricate large-scale structure instead of a featureless mass, and it did so while entropy climbed steadily the whole time.

Would you like the evidence? We can see the fingerprints of those early fluctuations in the Cosmic Microwave Background, usually shortened to CMB, which is the faint glow left over from roughly three hundred and eighty thousand years after the Big Bang. It is the oldest photograph in existence. Arno Penzias and Robert Wilson stumbled across it in 1964 as an inexplicable hiss in their radio antenna that they initially blamed on pigeon droppings. Later missions, COBE, WMAP and Planck, mapped tiny variations in temperature across the whole sky. Those variations are the seeds of everything that followed, and gravity was the partner that led entropy through the dance.

These earliest moments illustrate the Second Law of Thermodynamics rather beautifully, though the law is more often quoted than stated correctly. It is usually rendered as the claim that entropy will increase as time moves forward. What the law actually says is that entropy won't decrease. Those are different claims. The Second Law permits entropy to hold steady in ideal theoretical cases, and what it forbids is a spontaneous decrease in an isolated system. That distinction is going to matter a great deal in a few pages, so it's worth holding on to.

The early universe also poses an obvious question, which is what caused those quantum fluctuations in the first place. We will come to that in Chapter 8. For now let's follow the expansion and ask where it is going in such a hurry.

The universe is approximately 13.8 billion years old. A more careful statement is that the Big Bang is believed to have occurred 13.8 billion years ago from our vantage point here on Earth. That estimate rests on precise measurements of the CMB by missions including the Wilkinson Microwave Anisotropy Probe and the Planck satellite.

As for how it all ends, several scenarios are on the table.

The first is Heat Death, sometimes called the Big Freeze. The universe carries on expanding, stars exhaust their fuel, galaxies drift beyond each other's reach, and what remains is cold, dark and thin. The timescales here are difficult to hold in the mind. If protons decay, as some theories predict, ordinary matter disappears in something like 10^40 years, leaving only black holes. Even those evaporate eventually, over something in the region of 10^100 years. That figure has a name, because a one followed by a hundred zeros is a googol, so the last black holes wink out after roughly a googol years. If protons turn out to be stable, the timescales run very much longer still.

One thing to hold in mind alongside those figures. These numbers are still moving. A recent calculation that folded Hawking radiation into the lifespan of white dwarfs pulled one long-standing estimate down from 10^1100 years to 10^78, which is a revision of a scale that has no everyday equivalent. When you meet a number like this in a book, including this one, treat it as our best working figure for now, not as a fact.

The second scenario is the Big Rip. If the influence of dark energy strengthens over time, it could eventually overwhelm every other force, tearing apart galaxies, then stars and planets, then atoms themselves. Working from one particular model of how dark energy might behave, this has been placed at roughly twenty-two billion years from now, with the Earth coming apart about half an hour before the end. It remains highly speculative.

The third is the Big Crunch. If expansion reverses, whether through gravity or something else, the universe could collapse back into a hot dense state resembling the one it began in. In 2022 Cosmin Andrei, Anna Ijjas and Paul Steinhardt published a calculation suggesting this could begin surprisingly soon, with expansion ceasing in as little as sixty-five million years and contraction following within about a hundred million. That headline deserves its condition attached: their result depends on dark energy being a dynamic field known as quintessence and not the fixed cosmological constant of the standard model. If dark energy is constant, none of it applies.

The fourth I'll call the Infinity and Beyond Model, which is simply the idea that the universe expands forever and none of the above ever happens.

If the prospect of the universe ending unsettles you, remember that the universe may itself exist Somewhen, exactly as you and I do. Even after the moment of its destruction, every moment in which it existed would go on existing at its own location along the fourth dimension. The ending would be an edge, not an erasure.

If that hasn't done the job, the Reverse-Verse may offer something further.

This is where mirrored pictures of time start to feel natural instead of exotic. Rather than forcing entropy to run backward, which no physicist would accept, we could be looking at two flows of time that both carry increasing entropy away from a shared low-entropy middle. From our side, that middle looks like the Big Bang. From the other side, it looks like theirs.

Several established frameworks bear on this.

Conformal Cyclic Cosmology is the first of them. Roger Penrose's model proposes a cyclic universe that never passes through a Big Crunch. As the universe reaches maximum entropy in the remote future and loses every trace of meaningful structure, that infinite structureless state turns out to be mathematically identical to a low-entropy beginning. A new Big Bang emerges from the remains of the old universe, not because anything reversed, but because at that extreme the difference between the two states dissolves.

Loop Quantum Cosmology offers a second route, and this alternative, built on quantum gravity, replaces the singularity of the classical Big Bang with a bounce. Where the standard model demands a point of infinite density, Loop Quantum Cosmology suggests that quantum effects make gravity repulsive at extreme densities, so a previously contracting universe rebounds into an expanding one instead of collapsing to a point.

The Big Crunch is the third. In this scenario, expansion slows, stops and reverses under gravity, and everything compresses into a state of extreme density. It would be a mirror of the Big Bang, and the interesting question is what happens at maximum compression.

One of the strongest arguments for any cyclical model comes from the Second Law itself, first clearly formulated by Rudolf Clausius in the nineteenth century and among the most rigorously tested principles in all of physics. Recall the careful version: entropy can hold constant or increase, but it will never spontaneously decrease.

That raises a question worth taking seriously. What happens when entropy reaches its maximum possible state? In several cosmological scenarios, whether by heat death or by crunch, the universe may evolve toward a condition where no further increase in disorder is available and no usable energy remains to do any work. The Second Law isn't violated at that point. Entropy hasn't decreased; it has simply plateaued.

At that entropic ceiling the system sits in a kind of equilibrium, and this is where theorists have wondered whether such an equilibrium might function as a tipping point. The suggestion doesn't contradict thermodynamics at all. It proposes instead that at the ceiling something new becomes possible, not because entropy falls, but because the state of the universe has changed so completely that the old description no longer applies.

The Mirror Universe Hypothesis, or what we're calling the Reverse-Verse, offers an elegant possibility here. Once one universe reaches maximum entropy, a complementary universe is born or revealed, one in which entropy increases in the opposite direction of time.

Picture throwing a ball into the air. It leaves your hand fast and then slows. Gravity resists it the whole way up until it reaches the top of its arc, and at that moment the forces acting on it compel it to reverse direction. The ball falls back not because you threw it too weakly to rise, but because you threw it too weakly to overcome the counter-force that was pulling against it from the first instant.

Thermodynamics may behave in a similar way. When entropy reaches its limit, whether at heat death or at a crunch, the only motion left may be reversal, and rather than contradicting the Second Law this would follow from it.

Someone might ask what exactly is doing gravity's job in that picture, and it's a fair question. The ball turns around because gravity pulls on it the whole way up. Throw the same ball out in the deep space between galaxies and it just keeps going, because there's nothing pulling back. So if entropy behaves anything like the ball, something has to be playing gravity's part, and nobody has yet worked out what that might be.

I find that a wonderful question rather than a discouraging one, and it's exactly where Chapter 5 goes, with an idea of my own about what might be doing the pulling.

With that caveat standing, the conclusion is a powerful one. If entropy increases in our universe, and if time and entropy are linked, then two universes could exist in symmetry, one moving forward in time from the Big Bang and another moving backward from the Big Crunch.

On this view, what we call the beginning of our universe is one extreme of a larger cyclical system. At each end, whether Bang or Crunch, entropy is low, which means time has no clear preference there. It is only as the universe expands or contracts that entropy climbs and the sense of forward motion appears.

Notice that entropy never has to reverse in this picture. It can increase throughout. The trick lies entirely in where we stand. From inside the Reverse-Verse, time flows forward just as it does for us, and entropy increases just as it does for us. From a higher-dimensional vantage point capable of seeing both timelines at once, entropy increases away from the centre in both directions, like ripples spreading from a single stone.

Whether that central point is the Big Bang or the Big Crunch, or what I sometimes think of as the Big Tear, becomes almost beside the point. They may be the same event seen from opposite sides. The Second Law remains entirely intact, operating across two mirrored temporal axes and doing its ordinary job on each.

And perhaps, though I put this forward tentatively, it is the tension created by this two-way expansion of entropy that produces a kind of balancing system. If entropy is pulling outward from a shared centre along two timelines, that pull might itself constitute a constraint. I call this the Entropic Tension Force. It isn't a standard theory and no physicist has proposed it. It is an intuitive idea of my own, and Chapter 5 is where I make the case for it properly. For now, let's stay with the more established material.

So how do you fit into any of this?

Earlier we considered the possibility that all times exist at once. If that is right, then you exist at every moment of your life permanently. But let's pause that idea deliberately and assume the opposite, that Block Theory is wrong, that Somewhen is a pleasant fiction, and that now is a once-off state and the only time there is. If time then operates in a loop, what follows?

It would mean the film of the universe is played and replayed endlessly. Only the present exists, and there are long stretches in which you don't exist at all, but the moments in which you do live will play again and again. You must wait your turn, and yet you never experience the waiting, because there's no you present to experience it. From your point of view the showings run back to back with nothing in between.

If that is the case, then it isn't heat death or the Big Crunch or any other ending that ought to frighten you. It is the prospect of an infinite universe. A Reverse-Verse can't form from a universe that never reaches an end, because there is nothing for it to reverse from. In the introduction I suggested that the end of the universe might be not merely inevitable but preferable. Now you can see why.

Let us un-pause the first idea and set the two models side by side. Each theory can hold on its own, and in my view they can also work together. Whether all moments exist simultaneously or whether they queue up and take their turn, the conclusion lands in the same place: every moment may exist Somewhen. The only remaining question is whether all those moments are playing right now or waiting to be played again, and from the inside there's no way to tell the difference and no practical consequence if you could.

Either way, you're part of the history of time. Like every person, every rock and every tree, you've a role in that history that only you can play, and you'll play it, forwards and backwards, for as long as there is time in which to play it.

4. ENTERING THE COUNTER-VERSE

In the previous chapter we explored the Reverse-Verse, a model in which time eventually reaches a Big Crunch and then reverses, retracing every step it took on the way out. It is an elegant idea and it has the virtue of requiring only one universe. There are alternatives worth putting beside it, though, and this chapter deals with the most interesting of them.

What if, instead of time rewinding our own history, an entirely separate universe is born, one whose arrow of time runs opposite to ours while its inhabitants experience their own time as perfectly ordinary?

There is recent work that bears on this. A 2025 paper in Scientific Reports by Thomas Guff, Chintalpati Umashankar Shastry and Andrea Rocco examines what happens when a quantum system interacts with its environment, which is the situation almost every real system is in. Physicists routinely use an approximation in these cases, and it has long been suspected that the approximation itself quietly inserts a direction of time, which would make the arrow of time an artefact of our mathematics and not a feature of the world. The paper argues mathematically that this isn't what happens. The time-reversal symmetry survives the approximation, and the resulting equations don't prefer one direction over the other. That supports the theoretical plausibility of a Counter-Verse in which time flows opposite to ours and cause still reliably precedes effect.

So here is the idea. The Counter-Verse is a universe paired with our own, which begins moments after what we perceive as the Big Crunch and evolves toward what we consider the Big Bang. From our vantage point its timeline is inverted. If you could somehow stand outside both, you would see time in our universe moving one way and time in theirs moving the other.

Think of reality as a large clock on a wall, and an old-fashioned one with hands, not a digital display. Label nine o'clock as the Big Bang and three o'clock as the Big Crunch. From nine round to three, the hand sweeps across the top of the face, moving rightward from where you stand. That is our timeline, and it is the whole of our history.

Now imagine you could climb inside the clock and live on the hand itself, right at the tip. From your position the hand only ever moves forward. You have no sense of circling, because you can't see the face. Someone watching from outside the clock can see what you can't, which is that the hand is travelling a closed loop, sweeping rightward for half of it and leftward for the other half.

Here is where the two models part company, and the distinction is worth stating clearly because the whole chapter turns on it. In the Reverse-Verse, the hand reaches three o'clock and then comes back the way it came, retracing the same upper half of the face in reverse, passing through every position it has already occupied. In the Counter-Verse, the hand carries on from three o'clock the long way round, across the lower half of the face, through positions it has never visited before. One clock, two possible paths, and the difference between retracing your steps and continuing round is exactly the difference between the two ideas. The Reverse-Verse replays our history backwards. The Counter-Verse is a distinct universe with its own history, which dies before our universe is born and is born again at the end of ours.

Neither side perceives its own time as running backward. Each universe progresses forward within its own experience, and from a higher-dimensional vantage point each system comes from the other and leads back to it.

A crucial feature of the model is that nobody in the Counter-Verse thinks anything unusual is happening. Just as we remember the past and anticipate the future, so do they. Their future leads toward their Big Crunch, which is our Big Bang, and their past comes from our Big Crunch, or from whatever equivalent ending turns out to hold.

If we could look into their universe from ours, everything would appear to run in reverse. A star forming would look to us like a star collapsing and dissolving back into raw hydrogen. To them it is hydrogen gathering into a new star, which is what it is. Their galaxies spreading apart would look to us like a universe shrinking, while for them it is ordinary cosmic expansion.

Anyone old enough to remember cassette tapes has a ready image for this. While one side plays forward, the other side is being wound back. That is the shape of the relationship. The difference is that each moment has a counter-moment, and the music on the two sides isn't the same music. Each side is played forward, and each side has its own tunes.

It's worth sitting for a moment with how strange the relationship becomes when you follow it through. Their past comes from our future. The end of everything we'll ever be is, from their side, the deep and settled history that their equivalent of geologists would dig into. Every particle of the universe we're still in the process of using up is, to them, ancient material laid down before their first star. Meanwhile the moment they think of as their beginning is the moment we're all travelling toward, and which we regard as the end of everything. Two civilisations, if there were ever such things on both sides, would each be looking back on the other's future and calling it prehistory.

There's no privileged position from which to say which of those two accounts is correct, because both are correct within their own frame, and there's no third frame available to arbitrate. The vertigo you may be feeling is the appropriate response, and it doesn't go away with familiarity. What it does do, eventually, is loosen the grip of the assumption underneath it, which is that there must be one true direction and that we happen to be facing the right way.

The Counter-Verse therefore doesn't contradict the Second Law of Thermodynamics. It obeys it, in the opposite temporal direction from ours. Just as gravity allows an object to be thrown upward before pulling it back down, perhaps thermodynamics allows its limits to be approached, and tension to build, before the system moves the other way. In this model the Big Crunch and the Big Bang are the limit-testing moments, where for a fraction of a second uniformity approaches its extreme and then falls rapidly away as entropy climbs again.

I should repeat the caution I raised at the end of the last chapter, because it applies here with equal force. The thrown ball reverses because gravity acts on it continuously, and no equivalent restoring force is known for entropy. That gap hasn't closed just because we've changed models.

At first hearing, a Counter-Verse might sound like pure invention. Several existing frameworks in physics provide a foundation for it, and they're worth taking in turn.

The CPT-symmetric universe supplies the strongest argument, and it comes from work by Latham Boyle, Kieran Finn and Neil Turok, developed further in a 2022 paper by Boyle, Teuscher and Turok titled *The Big Bang as a Mirror: a Solution of the Strong CP Problem*. Apply the three symmetries of charge, parity and time to the universe as a whole, and a mirror-image universe emerges naturally as its counterpart. In their model our universe isn't alone. It has a time-reversed twin made of antimatter instead of matter, evolving toward what we call the Big Bang. Their Big Bang is our Big Crunch, which means their universe is born from what we would see as destruction.

The model also makes a concrete prediction about dark matter, and it's worth spelling out because it is the one place in this book where a speculative cosmology commits itself to something a laboratory could in principle find or fail to find.

Dark matter is the name we give to whatever is supplying about five sixths of the gravitational pull in the universe without emitting or absorbing light. We can see its effects in the way galaxies rotate and in how light bends around galaxy clusters, and we've never detected a particle of it. In the CPT-symmetric model, the candidate is a neutrino, which is worth a word of its own. Neutrinos are extraordinarily light particles that barely interact with anything; trillions of them are passing through your body as you read this and essentially none of them will touch a single atom of you. The neutrinos we know about all spin in one direction relative to their motion, which physicists call left-handed. The model proposes that right-handed neutrinos also exist, that they're very heavy rather than very light, and that one of the three is stable and has simply been sitting there since the beginning, supplying the missing gravity.

That is a real prediction with a number attached, not a gesture. Note where the dark matter sits in this picture. It isn't in the twin universe, reaching across to supply gravity on our side. It is here, in our own universe, in the form of a particle we haven't yet found.

There is one more thing about this model that I'd sooner put in front of you than tuck away, because it creates a difficulty for the book. The CPT-symmetric universe doesn't need cosmic inflation. Its authors present that as a feature, since it explains the smoothness of the early universe without invoking an extra epoch of exponential expansion. In Chapter 8 I'll make a fairly enthusiastic case for inflation, which has passed several demanding observational tests. Both of those can't be the final word. I'm not going to resolve that here, and I'm not going to quietly drop one of them either. This is exactly the situation an idea-system is built for: two useful models that can't both be wholly right, each held for what it explains, with the tension left visible.

Conformal Cyclic Cosmology offers a second foundation. Roger Penrose's model, which we met in the last chapter, proposes that at the end of an expanding universe, once particles have lost mass and structure has dissolved, the universe effectively resets and transitions smoothly into a new beginning. If entropy becomes meaningless at that extreme end state, it might create exactly the conditions for a new low-entropy Big Bang. That offers a framework in which a Counter-Verse could emerge without violating any thermodynamic principle.

Loop Quantum Cosmology supplies a third. It proposes that at the enormous densities reached in a Big Crunch, gravity turns repulsive because of quantum effects, which prevents a singularity from forming and instead produces a bounce into a new phase of expansion. Instead of one forward-expanding timeline, the universe may oscillate, each Big Bang following a prior Big Crunch. Most versions assume the same universe re-expands, but it takes no great stretch to imagine a variation in which the post-bounce universe runs on an inverted timeline relative to its predecessor, which would produce a Counter-Verse.

Black-hole cosmology is the fourth. In the 1970s Raj Kumar Pathria and the mathematician I. J. Good separately proposed that our universe might be the interior of a black hole, an idea also known as Schwarzschild cosmology. More recently Nikodem Popławski at the University of New Haven has extended this into a torsion-based model using Einstein-Cartan gravity, which avoids singularities altogether and produces a bounce into a new universe on the far side of a black hole's event horizon.

Not all of these require a new universe to run counter to its parent. What they share is that they allow a new universe to emerge from the end of an old one, and a Big Crunch would be an event of much the same character as a black hole collapsing. Put together, they make the Counter-Verse a possibility worth taking seriously. It also has the appeal of speaking to two open problems at once: what dark matter actually is, and what happens in the final moments of a dying universe.

Now for a question the book owes you, and which I would rather raise myself than leave you to notice.

I've now given you two models. In one, time reverses and retraces our history. In the other, a separate universe runs the other way with a history of its own. What observation would tell us which, if either, is right?

The honest answer is that almost nothing currently could. From inside our own universe, the two look very nearly identical, because in both cases we see what we already see, which is entropy increasing away from a low-entropy beginning. Neither model predicts anything we can go outside and check, because there's no outside available to us. A partner universe on the far side of the Big Bang isn't somewhere we can point a telescope.

That's no reason to drop either idea. It's a reason to hold them a little differently from the way we hold something like the age of the universe, which we've actually measured. Some will say that an idea we can't yet test isn't really physics at all, and they may have a point. But every idea worth having starts out untested, and the interesting question isn't whether we can check it today. It's what we'd need to look at if we ever could.

There is one avenue, and it is what the next chapter is about. Suppose two universes share a boundary, and suppose the state of one bears in some way on the state of the other. Any such relationship would have to leave a trace. And the only surviving record of our universe's earliest moments is the Cosmic Microwave Background.

That is where I would go looking, and Chapter 5 is my attempt to say what I think we might find.

So what does any of this mean for us?

If the Counter-Verse is real, it changes the shape of an ending. It suggests that every moment belongs to an eternal cycle, that our existence is finite within each turn of that cycle, and that the greater system isn't finite at all. We may never experience the Counter-Verse in any sense. Knowing it might be there still supplies a certain balance, because it means our universe has a counterpart and the sequence of creation and destruction has no last term.

Most of all, the model suggests that while the universe is in constant change, the underlying structure of reality is stable and repeating. Our universe wouldn't end so much as wait its turn to be born again.

It would mean that when you take your last breath, you're in a sense waiting, thankfully without awareness and without discomfort, to take your next one.

The ordinary picture, in which the universe runs down once and stops and that is the whole of it, isn't the only picture consistent with the physics. It has a competitor, and the competitor isn't mysticism. It comes out of symmetry principles that were established for entirely unrelated reasons and that have survived a great deal of testing.

None of this contradicts or cancels the idea of Somewhen. Like the hours marked on a clock face, every second and every minute exists in both universes at all times. At any given moment the hand is pointing at every hour, and on the tip of that hand the countless lifeforms inside the clock perceive their past and their future and infer from the difference a sense of motion that they call time.

The Counter-Verse doesn't contradict Somewhen, and it doesn't require it either. Both offer something worth sitting with.

5. ENTROPIC TENSION FORCE

Twice now I've made you a promise and deferred it. In Chapter 3, I said that the image of a ball thrown into the air only works if something plays gravity's role for entropy, providing a real restoring force rather than a poetic one, and that physics knows of no such thing. In Chapter 4, I said that the Reverse-Verse and the Counter-Verse can't currently be told apart by any measurement, and that the one place worth looking for a trace would be the oldest surviving record of our universe's earliest moments. In both cases I pointed at this chapter.

So here it is. What follows is an idea of my own, offered as speculation and labelled as such throughout. I'm not a physicist, this hasn't been through peer review, and no working cosmologist has proposed it. I call it the Entropic Tension Force, or ETF for short, and I want to be precise about what kind of thing it is before I start arguing for it.

A piece of housekeeping comes first. In Chapter 3 we did the work on what entropy actually is. It's a measure of how many ways the parts of a system can be shuffled while still looking the same from outside. We also covered why a smooth early universe counts as low entropy once gravity is in the room. I won't repeat any of that. But there is a related confusion I do need to clear up, because almost everyone arrives with it.

In everyday speech, order sounds tidy and disorder sounds messy, so a far-future universe that is cold, thin and empty sounds like the tidiest thing imaginable. It is the opposite. A heat death isn't low entropy. It is maximum entropy: everything spread thin, everything the same temperature, no free energy left anywhere to do anything interesting. The universe wouldn't have become neat. It would have become so thoroughly shuffled that no further shuffling makes any difference.

That matters because it rules out a tempting argument. You can't argue that infinite expansion must eventually reverse because it would otherwise produce a low-entropy final state and violate the Second Law. It produces a high-entropy final state, and the Second Law is entirely satisfied. If there is a reason to think expansion turns around, it has to be a different reason.

Many observations suggest that the expansion of the universe isn't merely continuing but accelerating, and we give the name dark energy to whatever is driving that. If acceleration keeps winning indefinitely, we get the Big Freeze picture: stars burn out, black holes evaporate through Hawking radiation, and what remains is a dilute bath of particles and faint light spread across an unimaginable volume. No law is broken anywhere in that story. Entropy has climbed steadily the whole way, exactly as it should.

Which leaves a structural question, not a thermodynamic one. Do the rules of the cosmos really prefer a one-way leak into featureless stillness? Or is runaway expansion a phase and not the last word?

Picture two identical robots tied together by a long elastic cord, sprinting in opposite directions down a track. As the cord stretches, one of two things happens. Either the cord snaps and each runner coasts away forever, or the elastic holds, slows them both, and hauls them back.

The second outcome contains the feature that matters, and it is the feature people usually miss. It isn't simply that the elastic prevents infinite separation. It is that the force exerted by one runner is what pulls the other one back. Each runner's motion isn't merely complementary to the other's; it is necessary for it. Neither could be doing what it is doing without the other doing the same thing in the opposite direction. Take one runner away and the cord goes slack and the whole arrangement collapses.

That mutual necessity is the whole of what I'm proposing. In the cosmic version, the runners are two time-oriented halves of a larger whole, our universe and its partner. As each side races away from a shared low-entropy boundary, which is what we call the Big Bang, entropy increases on both sides. The stretch builds something across the system as a whole. If the cord is real, separation doesn't grow without limit; the dynamics are eventually pulled through a turning point, whether that turns out to be a crunch, a bounce or a reset, and no thermodynamic rule is broken anywhere in the process.

One thing about that arrangement needs stating clearly, because the whole idea stands or falls on it.

ETF isn't a signal, and it isn't a force in the sense that electromagnetism is a force. Nothing travels from one side to the other. It couldn't, because travel requires a direction to travel in, and the shared boundary is where both arrows of time begin. What ETF describes is a constraint on the joint state of the two universes. Consider a simpler case. If I tell you that two numbers must add to ten, then fixing one fixes the other, and no message has travelled between them. There's no signal, no delay, no direction. There is a relationship that both must satisfy. The correlation is real and it has consequences, and yet nothing has moved.

My suggestion is that the two halves of a mirrored cosmos might be related in something like that way at their shared boundary, and that this relationship, and not any transmission, is what would leave the pattern I'm about to describe. Whether such a constraint is possible, and what could impose it, I can't tell you. It is a conjecture. But it is at least the right shape of conjecture, and I would rather present a speculative idea in a coherent form than in a form that sounds more dramatic and can't survive its first question.

Here is the speculative move itself.

Imagine two universes sharing a common low-entropy boundary. On our side, time as we experience it runs away from that boundary. On theirs, it runs the other way and is always forward for them. Near their end, which is our opposite beginning, their cosmos is in a state of extreme gravitational complexity, with black holes merging, structures collapsing and horizons distorting, and entropy high and still climbing. Suppose now that the joint system has to satisfy certain smoothness conditions at the shared boundary. Then the pattern of their late-time entropic wrinkles and the pattern of our earliest irregularities wouldn't be independent of one another. They would be two faces of a single constrained state, and our side would carry a very faint bias in where matter first began to condense.

That cross-boundary relationship is what I'm calling the Entropic Tension Force. It wouldn't replace the standard account, where quantum fluctuations get stretched to cosmic scale by inflation. It would add the faintest tilt on top of it, far too slight to see in any one place and just possibly visible in the pattern of the whole.

What would that look like from where we sit? If ETF is real, it should leave statistical fingerprints in places we already study closely. There are three I would expect.

The first is mild large-scale asymmetries or alignments in the Cosmic Microwave Background, patterns that ought not to have any preferred direction and yet appear to. The second is small correlated departures from perfect randomness in the primordial fluctuations, structure where the standard picture predicts only noise. The third is regions of the universe, or early structures within it, that look older or more developed than the age of the universe leads us to expect.

All three of those describe things that have actually been reported. Let me be careful about what that does and doesn't mean, because this is exactly where a speculative idea can be made to look far better supported than it is.

There are genuine anomalies in the CMB. Certain large-scale features do appear to align in ways that a perfectly isotropic universe shouldn't produce, and the pattern has been discussed for two decades, sometimes under the unfortunate nickname of the axis of evil. There is also live discussion about galaxies observed by the James Webb Space Telescope that appear more mature than the standard timeline comfortably allows, and about a reported preference in the direction galaxies rotate.

Most cosmologists think the CMB anomalies are probably flukes, and the reasoning behind that is worth knowing. We have exactly one sky. Look at a single set of data closely enough, test it against enough possible patterns, and something will eventually clear the bar by chance, with no second sky available to check it against. The Webb results have conventional explanations in play too, including revisions to how we weigh very distant galaxies and how we model the earliest star formation.

So these are open questions, not puzzles waiting for my solution. The claim I can make is narrower. If ETF were real, it would produce effects of roughly this character in roughly these places, and it is at least interesting that the places where the standard picture is under mild strain are the places the idea points to.

Which raises the question my last chapter taught you to ask, and it would be poor form to teach you a test and then exempt myself from it. What would show that I'm wrong?

The answer is clean enough. ETF has one thing going for it observationally, and that's the anomalies. If they resolve into noise as our measurements improve, and the Webb results turn out to have ordinary explanations, then ETF has nothing left to point at. At that point I'd put it back on the shelf, and I'd like you to hold me to that.

That isn't modesty. It's the whole difference between an idea and a belief. I've spent this book noting how easily clever people end up defending a theory the way you'd defend a house, and it would be a poor sort of hypocrisy to exempt my own.

I should also say what ETF would need in order to become more than a conjecture, since I've already ruled out putting mathematics in this book. It would need a formal definition tying it to entropy, gravity and time dilation. It would need a way of writing down how entropy on one side relates to entropy on the other. It would need a change to the equations that govern how the universe expands. And it would need to predict a number that differs from the standard model by enough for somebody to go and measure. Knowing what a theory would require isn't the same as having one, and the gap between the two is where most interesting-sounding ideas quietly die.

The idea of cyclical universes and interconnected forces of creation and destruction isn't remotely new, and I find that heartening. Many ancient cultures pictured time as a self-consuming cycle. The Ouroboros, a serpent swallowing its own tail, appears in Norse mythology, in Egyptian cosmology, in Greek alchemy and in Celtic paganism, and in each case it stands for the eternal return of existence.

Those symbols weren't physics and were never intended as physics. They may, though, have been an intuitive reach toward something that our equations are only now circling: that time might be a looped and interconnected system instead of a straight line, in which the forces of one epoch bear on the next.

If ETF is real, then four things follow. The extreme entropy of a dying universe would directly shape the initial conditions of a new one. The distribution of mass after a Big Bang would carry a faint bias from its partner. Variations in gravity and in the rate at which time passes across cosmic structures might be residues of that bias. And our universe wouldn't be an isolated system at all, but one participant in a larger interconnected cycle of creation and destruction.

I offer all of this alongside the established material, not in place of it, and I offer it as an illustration of how a flexible idea-system works in practice. You can slot a speculative idea in beside well-tested ones, see what it would explain if true, keep track of what would count against it, and decline to marry it. ETF is my demonstration of that method using my own idea, which is the only fair way to demonstrate it, since it is easy to be rigorous about other people's proposals and considerably harder to be rigorous about your own.

There is one implication I do want to draw out, because it's the reason I keep coming back to this idea, even if it wasn't the reason I first had it.

If the state of one universe at its end bears on the state of its partner at the beginning, then the arrangement of everything that exists at the close of a cosmic era isn't simply discarded. It is the raw material of the conditions that follow. Every star that formed, every galaxy that collapsed, every black hole that swallowed what it swallowed would contribute to the pattern that a subsequent universe inherits, and in a truly cyclical arrangement that pattern eventually circles round to us again. On that reading we aren't tenants in a building that will one day be demolished. We are participants in something that carries forward, and the distinction isn't sentimental. It is the difference between a system that discards its own history and one that uses it.

Our universe, on this reading, isn't an isolated event with a beginning and an end. It is one side of a larger and continuing story. What that means for how you spend a Tuesday is a question I'll come back to in Chapter 11, once we've the rest of the structure in place.

Before we leave this chapter, there is an alternative worth putting beside it, because ETF isn't the only route to a reset.

Consider the Big Rip again. We met it in Chapter 3, where it was placed at roughly twenty-two billion years from now under one particular model of dark energy. In that scenario, dark energy doesn't merely persist but strengthens, until it overwhelms every other force in turn. Galaxies are pulled apart, then solar systems, then planets, then atoms, and finally the constituents of atoms. What I want to add here is what the end state actually resembles. Once matter has been broken down into fundamental particles and radiation, the universe is left in a hot, dense, high-energy condition that isn't at all unlike the one it began in. The ending has, without anybody intending it, reproduced the beginning.

Could a new universe emerge from that? Not by reversing, and not by any partner reaching across a boundary, but simply by bouncing ever forward into another beginning. If so, then even without ETF, and even without a Reverse-Verse or a Counter-Verse, the cosmos might still be embedded in a perpetual cycle of death and rebirth. Each iteration would unfold in what looked from inside like a single direction of time, while from a wider vantage the whole sequence would form a loop, or at least a chain with no last link.

And if such a cycle persists, whatever its shape, then the question that follows is the one I find hardest to stop thinking about. What unimaginable universes and worlds might be born next?

To take that seriously we need to move beyond the endless loop and beyond the four dimensions we've been working in. Next, the fifth dimension and the Multi-Verse, where a similar but different version of you decided not to cut their own hair the night before that school photograph was taken.

6. MULTI-VERSAL THINKING

So far we've stretched our thinking about space, time and self across four dimensions. We've looked at time looping, time reversing, and all of time existing at once. Every one of those pictures has kept us inside a single system where everything that has happened, is happening and will happen plays out the same way every time. Our model has been four-dimensional throughout.

The good news is that's probably it for you. Or rather, for this version of you. That's not to say you can't get some joy out of a little fifth-dimensional thinking.

Earlier, when I introduced the Reverse-Verse and the Counter-Verse, I was careful to say we weren't talking about universes where a different version of you was running around making different choices. We were talking about the flow of time repeating or mirroring this exact reality. But that doesn't mean the other sort of universe isn't out there too. It just means we hadn't got to it yet.

So what is the fifth dimension?

Before I answer that, I owe you something. Back in Chapter 2 I said I'd be straight with you about how I'm using dimensional language, and this is where that debt comes due.

You'll often see the fifth dimension described as though string theory hands it to us, ready-made, as a place where parallel versions of our universe sit side by side. It doesn't. String theory and its broader framework, M-theory, do propose extra dimensions, ten or eleven depending on the model. But those extra dimensions are compactified, which is the garden hose again: curled up so small that we never encounter them, real but far below anything we could notice. They aren't a filing cabinet of alternative worlds. They're part of the fine structure of the space we're already standing in.

What I'm about to build is different, and I'd rather you knew that from the start. It's a ladder for thinking, a way of organising kinds of possibility so we can talk about them without tripping over ourselves. The fifth rung is where outcomes vary. Later rungs are where the rules vary, and then where the rules about the rules vary. It's a useful scheme and I'll use it for the next four chapters, but it isn't a claim about the geometry of spacetime and no physicist would recognise it as one.

Why be so fussy about this? Because the ladder doesn't need string theory's endorsement, and pretending otherwise would be borrowing authority I haven't earned. The idea stands or falls on whether it helps you think, which is a fair test and one I'm happy to be judged by.

With that said, here's the rung we're standing on.

The fifth dimension isn't up, or down, or another timeline. It's a way of talking about shifts between entirely separate realities: parallel universes running on the same physical laws, but with different outcomes. These aren't fantasy realms full of flying dragons, necessarily. Our closest neighbours would be almost identical to us, differing only in that one coin flip landed the other way.

Think about all the variable events happening across the Earth in the next hour. Now expand that across the solar system, then across the whole universe. If every possible outcome can spawn its own universe, you can see how quickly the combinations run away from you.

Several models support something along these lines.

String theory and M-theory unify particles and forces by requiring more than four dimensions, and some versions of the theory do allow for a much larger structure containing many universes with different properties. That larger structure is where the ladder's fifth rung finds its nearest physical relative.

The Many-Worlds Interpretation was proposed by Hugh Everett in 1957. In quantum mechanics, an unmeasured particle exists in a spread of possible states, and the standard account says that on measurement the spread collapses to a single outcome. Everett's suggestion was that nothing collapses at all. Instead the whole spread persists, and every possible result happens, each in its own branch. **Decoherence**, the loss of quantum interference when a system meets its surroundings, is what makes those branches effectively unable to communicate. You can't see the other outcome, and you can't reach it.

One point about Many-Worlds is worth having straight, because it's regularly misreported. It's an interpretation of quantum mechanics, not a rival to it. It makes exactly the same predictions as every other interpretation, which is precisely why physicists have been arguing about it for nearly seventy years without settling it. Nobody can build an experiment to tell them apart. It's a disagreement about what the mathematics means, and the mathematics is not in dispute.

Brane cosmology offers a third picture. In this version of string theory our universe is a four-dimensional membrane, a "brane", floating in higher-dimensional space. Other branes could sit nearby, each potentially hosting a universe of its own.

None of these are proven. All of them are mathematically consistent and widely explored by physicists trying to understand why our universe has the properties it does.

Now picture a tree, branches splitting into more branches, endlessly. Universes nearer to ours on the tree are more like ours. On a nearby branch you'd find a version of yourself who wore different shoes today. Further out, one who chose a different career. Further still and humanity never evolved, and dinosaurs or something else entirely became dominant. Go far enough and Earth never formed at all.

The one thing every branch shares is the trunk. The same physics applies throughout. Time flows forward, gravity pulls masses together, entropy climbs. At the fifth dimension the rules hold steady even as the story changes. That's the whole point of this rung, and it's what separates it from the ones we'll meet later.

That raises an obvious question. Could we ever travel through it?

Time travel, as the films do it, tends to collapse under its own paradoxes. In Back to the Future, Marty travels into his past and changes it in ways that remake his present. But if Marty had really travelled to his own past, the changes he made couldn't have been changes. For Marty from Timeline A to go into the past of Timeline A, he'd have needed to have already done it. Otherwise he isn't visiting his past. He's creating a different timeline.

Perhaps a dual flow of time would let him make the journey, but only if his making it was already baked into his own history, with effect somehow predating cause. Even then, he couldn't go back and alter his own timeline, because the act of altering it would mean it was no longer his. His timeline would cease to exist, so he could never have set off from it in the first place.

Fifth-dimensional travel doesn't tie itself in the same knots. If each universe already exists, then travelling to another version of reality isn't altering your own history. It's stepping into an already-existing one that had nothing to do with your past until you arrived.

Practically, this is still deep science fiction. We've no known way to hop branches and there are good reasons to think nature keeps them firmly apart. Logically, though, it doesn't tangle itself the way time travel does, and I find that difference interesting in its own right.

Could we ever actually do it? That's Hollywood's problem for now, and I'll leave it to the Avengers. If it were possible, it would mean browsing the tree, hopping from branch to branch through alternative histories. It would also be spectacularly dangerous. You might arrive in a reality where Earth is a nuclear wasteland, or where life evolved to tolerate far higher temperatures, or where some more aggressive airborne bacterium dissolves you within seconds. The risks would be severe, to put it mildly.

If the fifth dimension is real, then something else follows. Each of those parallel universes might be paired with its own Reverse-Verse or

Counter-Verse. The structure we've spent four chapters building may not be unique to us at all. It may be one branch on an infinitely branching tree.

Each of those universes could be as predestined as our own, running the same physics with different starting conditions or small changes that widen over time. There may be a finite number of them, or an infinite number where everything that could happen does.

In Chapter 4 we touched on how a universe might be born from the death of another, and on the idea that a sub-universe might spawn from a black hole. Let me pick that thread up again, because it gives us a physical mechanism rather than just a picture.

The idea is that black holes aren't only endings. They may be beginnings, connected to what they create by something called an **Einstein-Rosen Bridge**.

Black-hole cosmology was first proposed in the 1970s by Raj Kumar Pathria and the mathematician I. J. Good, and it's sometimes called Schwarzschild cosmology. More recently the physicist Nikodem Popławski has revived and extended it using Einstein-Cartan gravity.

In his version, the enormous density at a black hole's core doesn't collapse to an eternal singularity. Torsion, a kind of twist in spacetime that arises from the spin of matter, halts the collapse at a finite but extreme density, and the whole thing rebounds. That rebound channels through an Einstein-Rosen Bridge, more commonly called a wormhole, and out into a new, closed baby universe.

The bridge itself was proposed in 1935 by Albert Einstein and Nathan Rosen. It's a mathematical solution to Einstein's field equations connecting two distant points in spacetime. It emerged from the same solution that describes non-rotating black holes, and it suggested a kind of tunnel: between two black holes, or between two universes.

So torsion prevents the singularity, matter inside the black hole never becomes infinitely dense, and instead of ending it takes a new path. It passes through a one-way tunnel and forms a fresh universe on the far side of the horizon.

What I like about this model is that it gives universes a way to branch from real objects we can actually point a telescope at, rather than from mathematics alone. And if the parent black hole is rotating, that rebound might imprint a preferred axis on the newborn universe, which would be a fingerprint of sorts. There has been interest recently in a possible preference in the direction galaxies rotate, though as with the anomalies we looked at in the last chapter, it's early and contested and I'd not lean on it.

Weave in the Entropic Tension Force from the last chapter and the picture gets denser still. If coupled universes share constraints at a low-entropy boundary, then late-time complexity on one side might tilt the earliest ripples on the other, producing tiny preferences in where matter gathers after inflation. It wouldn't replace the standard account. It would sit faintly on top of it.

Put plainly: our universe may have spawned whole new realities out of collapsing stars, and those same realities might, through ETF or something like it, quietly shape how our own universe formed. And our universe could itself be one such sub-universe, born out of something else's collapsing star. Not isolated canvases, then, but an interconnected tapestry.

All of which offers a rather good answer to one of the oldest questions we have. Why are we here?

Notice the question smuggles in an assumption. It presumes there's a special reason things happen, and some will behind it deciding which things.

If ours were the only universe, the odds of everything lining up for life to form on Earth, and then for you specifically to be born, would be vanishingly small. One universe, exquisitely tuned for life, looks like a loaded coin. A loaded coin invites you to ask who loaded it.

In a Multi-Verse the burden shifts entirely. With countless branches, some will happen to grow the conditions for life, and the ones that don't have nobody in them to notice. Our existence stops being a lottery miracle and becomes a selection effect. We're observing from a branch compatible with observers, because those are the only branches capable of containing anyone to do the observing.

Monarchs used to believe, and a few probably still do, that God chose them to rule. The reasoning was neat enough: they inherited the crown, so the crown must have been intended for them. Divine right doesn't survive a Multi-Verse, where in a great many worlds the monarch is never born, or is born a pauper, or Napoleon wins at Waterloo and reshapes the map, or hamsters evolved and rule over the lot of us.

That's the beautiful thing about infinity. If every possible version of our universe exists, then you were always going to be born and live exactly the life you've lived, in at least one of them. So was I. So was everyone. You were always doomed to spill coffee down your shirt on the way to that interview, at least in one universe. You were always going to meet, love and lose the people you did in this reality, at least in one universe. If you could play every possible combination in a lottery, you'd win the jackpot at every draw. Infinity lets you try every scenario, and the only entry requirement is that the scenario be possible at all, however unlikely.

Some people find that idea deflating. They'd rather be the ultimate expression of the universe's singular will to deliver them into existence, and they'd rather not hear that they're probably not even the best available version of themselves. Others find it a relief. I'm firmly in the second camp and I find it liberating.

For every universe where I fail, there's another where I succeed. For every heartbreak in this life, another where the love holds. I'll never be an astronaut here, but I probably am one somewhere. Maybe I got luckier earlier in life in some other reality, and maybe in this one I'll experience something unique and equally worth having precisely because of what went wrong here. In an infinite universe, not everything is possible. But everything possible is.

Now, if you've been following the branching argument closely, you may have arrived at a question of your own. I'd rather meet it head on than leave it sitting there.

If every outcome happens, and if there's always some branch in which you survive whatever's just befallen you, then from the inside you'd never experience not surviving. There's no branch in which you notice your own absence. So wouldn't you simply go on, forever, always finding yourself in the branch where you made it?

The idea has a name. It's called quantum immortality, and given the title on the cover of this book you're entitled to hear what I make of it.

The logic is cleaner than it first sounds. Death is the one event you can't observe from the inside. In every branch where you survive, you're there to notice. In every branch where you don't, there's nobody home. So your continuous experience, threading through the branches, would always land on survival, not because survival is likely but because it's the only kind of outcome your experience can contain. Follow that through and you'd never have the experience of ending. You'd only ever have the experience of an increasingly improbable continuing.

It's an arresting thought and I understand why it grips people. I want to lay out where it takes you, though, because I think the destination is more interesting than the argument.

The branches in which you continue after something serious are not the branches anyone would pick. If a person survives an event that kills them almost everywhere else, the version that survives is rarely the version that walks away whole. Nothing in the reasoning steers you toward the good outcomes. It steers you toward any outcome at all in which experience carries on, and those become stranger and more diminished the further you push it. Quantum immortality, taken at its word, isn't a promise of endless life. It's a promise of endless narrowing.

There's a second thing worth noticing. The argument, if it works, works for everyone and always has. It would apply to every person who ever lived and to every animal capable of experience. Whatever it offers, it isn't a discovery about you, and it certainly isn't a plan.

So I don't hold it, and I'd rather not have it attached to me. But I'd also rather not throw it away, because it points at something true that the rest of this book has been circling. Your experience really does have a shape from the inside that it doesn't have from the outside. You have never experienced a moment of your own non-existence and you never will, and that isn't a trick of the branching. It's true in a single universe with no branches at all. It was true before you were born and it will be true afterwards.

That, to me, is the more durable version of the same insight, and it doesn't require you to bet on anything. It's what Chapter 2 was reaching for with **Somewhen**, and it's what I'll come back to in Chapter 11 under a different name. Every moment you occupy is a moment you occupy. Nothing that has happened stops having happened. Whether the branches are real or not, that much holds.

I'd rather have that than an argument that survives me into progressively worse afternoons.

While the Multi-Verse sits comfortably alongside everything else we've explored, it doesn't have to. The point of this book isn't to build one tower where every floor depends on the one below. It's to set a lot of ideas beside each other and see what happens where they touch.

So let's do something drastic. Assume Chapters 2 to 5 are all nonsense. No Block Theory, no Somewhen, no mirrors, no ETF. Time moves forward in one direction only, for us and for any hypothetical being watching from outside.

Now imagine our universe ends, by Big Crunch or heat death, and a new one forms. And that this goes on, universe after universe, forever.

Here's what falls out of that. In an infinite sequence you don't need mirrors or branches to get repetition. Physics allows a finite alphabet of structures, and with unlimited attempts every arrangement eventually recurs, not because time loops but because the combinations run out and start over.

So even in a strictly forward-only cosmos, across unimaginable stretches, you get your exact history appearing again on some run. You get near-twins diverging in a single detail. You get worlds nothing like this one. From where you're sitting, that looks a great deal like the Multi-Verse we've just spent a chapter on.

Which is the beauty of an idea-system. Rather than needing one idea to hold, we let the alternatives sit together and look at what they share. In one model, all time exists and repeats across a two-universe loop. In the other, each universe's death sparks a fresh one, and eventually the same histories come round again. From your seat and mine, the effect is identical. Our time repeats.

That's infinity for you. Given enough of it, even a dead universe starts refilling with energy, because energy is never created or destroyed, only moved from one form into another. The total in a closed system holds steady no matter how long you wait.

Now, earlier we worried that free will might not survive Block Theory, if every somewhen already exists. Here's the strange thing. Layer the fifth dimension on top and the worry dissolves. Picture every point where a different outcome is possible as a branch splitting in two.

If you're young as you read this, it means every decision available to you isn't merely still possible. You, or different future versions of you, will make all of them. You say yes to everything. You say no to everything. You aren't the past-self of one future you. You're the shared past-self of every possible future you branching out from this moment.

It also means every future of yours that's possible will happen somewhere. In one reality you play the lottery and win. In a great many more you play and lose. Every outcome that can happen does, once it's truly possible. Some are difficult, some are wildly improbable, and they'll still exist Somewhen and somewhere along the branches spreading out from here. As we get older those options narrow, because less remains possible in the time left. What we can do is try to choose well, and steer toward the future we'd most want to find ourselves in.

Reality does us a kindness in hiding the future. If we could see every branch laid out ahead, we'd be paralysed by the smallest decision, or we'd only ever choose the preferred path and the whole branching structure would collapse into a single line. Instead we get one step, and then the next one.

I've talked about how the Multi-Verse dismantles the idea that we're chosen or destined for one path. It offers something else too, and this part matters most to anyone reading at a dead end.

If you ever feel you've no options left, remember that in an infinite universe there's a version of you who gets through this, and not only survives it but thrives afterwards. If it's possible at all, however unlikely, then it happens somewhere.

When I'm overwhelmed, I picture the future me who came out the other side. Not a different person, and not a luckier one. A version of me who started exactly where I am today. Then I ask what they might have done to end up where I want to be, and what small steps they might have tried.

That reflection guarantees nothing. But my odds improve against the version of me who never bothered to ask. Maybe the ideas that surface are long shots. Maybe this is the reality where the long shot lands.

Sometimes the smallest steps begin the greatest journeys.

7. THE STABILITY PARADOX

We've just come back down to a very personal scale, choosing the next single step in front of us. Let's zoom back out, because the next four steps in this book are the widest we'll take.

Before we climb any further, we need a handful of new terms. Without them the next few pages would turn into a tangle, so let's build the vocabulary first and use it after.

Here's the shape of where we're headed. Each rung from here changes something different. The fifth changes outcomes. The sixth changes the rules that produce those outcomes. The seventh changes the starting conditions those rules act on. The eighth changes what a rule is even allowed to be. By the time we reach the ninth, we're no longer looking at one system at all. We're looking at the whole collection of them, and asking whether ours is a lucky exception or something closer to the norm.

So far we've used **Multi-Verse** to describe the fifth dimension: a system of parallel universes where the laws of physics stay fixed and only the outcomes differ. These are the branches of a single cosmic tree. Every version of our reality plays out somewhere and somewhen on those branches. In one, you never get the job. In another, you invent the toaster. In another, you were never born.

But in every one of those universes, gravity still pulls, light still travels at 299,792,458 metres per second, and entropy never spontaneously decreases. They all run on the same rulebook. What changes is the result, not the rules.

What we're stepping into now is stranger. It's the **Sixth Dimension**, and from there we can compare entirely different Multi-Verses, each with its own rulebook, but still sitting inside a familiar overarching framework: cause still precedes effect, conservation laws are still tied to symmetries, quantum mechanics still supplies the backdrop. These systems aren't just branches of one tree. They're different trees, grown from different seeds.

Some trees share the same seed, meaning the same kind of physics, but grow under different conditions. Picture a universe where gravity is slightly stronger everywhere, or where the fine-structure constant is a touch larger so chemistry binds together a little differently. Stars, galaxies and maybe even life could still emerge, but the odds, the timescales and the recipes would all shift.

Push it further and you can imagine a whole family of such universes, each running the same equations with the dial on one constant turned up or down a notch. Most of that family would still look broadly recognisable. A little more gravity and stars burn faster and die younger. A little less and galaxies never quite pull themselves together at all. You don't need to change the rulebook to get a very different history. You only need to change one number in it.

The **Seventh Dimension** lets us vary something bigger than the constants. It lets us change the global template a universe starts from: its boundary conditions, its initial roughness, its large-scale geometry, even whether it has a single beginning at all, or a bounce, or a repeating cycle. Say our universe really did arise from a Big Crunch in a loop. A seventh-dimensional comparison might show you a cousin universe that begins in a quantum bounce instead, with much larger primordial ripples, so that black holes form earlier and far more often. Same species of tree, but this one sprouted in harsher soil, with less light and stronger winds.

Notice what's actually changed there. It isn't a constant this time. It's the shape of the story before the story starts, the difference between a universe that begins once and a universe that keeps beginning.

At the **Eighth Dimension**, things get properly strange. Here we're not just changing constants or starting conditions. We're changing the meta-laws themselves, the rules about what a rule is even allowed to do. Time might run in two directions locally, advancing and reversing like a video under a fidgety thumb, while still obeying some internal consistency of its own. Gravity might repel rather than attract under certain symmetries. Conservation laws might stop following the link between symmetry and conservation that the mathematician Emmy Noether proved in 1918, the same link that guarantees energy, momentum and charge are conserved in our universe because of deeper symmetries in nature. Most worlds built this way would never form stable structure at all, and life as we'd recognise it almost certainly wouldn't evolve in them. Think of it as different kinds of vegetation entirely: a stunted cactus surviving on a salt flat and an oak thriving in a rainforest are both plants, but they aren't playing the same game of growth.

Try to picture a world where cause and effect only hold for part of the time, switching on and off according to some rule we've no equivalent for. Most such worlds would fall apart before they'd properly started, structure collapsing as fast as it formed, or never forming at all. That's the eighth dimension's whole character. It isn't extreme by our standards. It's extreme by any standard, ours included, because ours is just one of the arrangements where the lights happen to stay on.

So here's where the four rungs leave us. The fifth dimension keeps the same rules and only the outcomes change: branches of our tree. The sixth keeps the same meta-framework but lets the rules themselves change: different trees, same ecosystem.

The seventh changes the rules and the starting template together: different trees, different climates. The eighth changes the meta-frameworks that generate the rules in the first place: different kinds of growth altogether.

Now that we've met the trees, we can jump to the **Ninth Dimension**, which is the forest itself, the space that connects all of these eighth-dimensional collections of possibility.

To keep the language tidy, we'll call each of these higher-order collections an **Alt-Multi-Verse**. Within any given Alt-Multi-Verse there can still be countless parallel universes branching fifth-dimensionally, but every branch on that particular tree shares that tree's rulebook.

In one Alt-Multi-Verse, gravity might repel instead of attract. In another, energy might not be conserved at all, because time-translation symmetry simply doesn't hold there. Some Alt-Multi-Verses would be incoherent, collapsing before any structure had the chance to form. Others might harbour strange, lawful orders we can barely imagine from in here, places every bit as internally consistent as our own but built on a foundation that shares almost nothing with it. We just don't know, and there's a real chance we never will. Nothing we've built, not a telescope, not an equation, has ever shown us past the edge of our own tree.

With those terms in place, we can ask one of the more important questions in this book. Do we live in a stable Alt-Multi-Verse? And if we do, is it likely to be infinite?

Modern physics keeps finding that the numbers describing our world are spookily well-tuned. Nudge the strength of the strong nuclear force too far in one direction and only hydrogen ever forms, with nothing heavier ever cooking inside a star. Nudge it too far the other way and every star burns through its fuel almost as soon as it lights, and nothing has time to become a planet, let alone a person. Nudge dark energy and galaxies never form, or everything collapses back in on itself too soon.

The starkest example might be the cosmological constant, the number behind the accelerating expansion we met in Chapter 3. Work it out from quantum field theory and you get a value around 120 orders of magnitude larger than what we actually observe. That's not a rounding error. That's the difference between predicting something should weigh a few pounds and getting an answer with over a hundred extra zeros tacked onto the end. Nobody has a fully satisfying account of why the real number is so much smaller than the theory expects, and it remains one of the most uncomfortable open problems in physics.

Physicists have a name for this uncomfortable pattern generally. They call it the **Fine-Tuning Problem**, and it's the observation that an awful lot of independent constants seem to sit exactly where they need to for anything complex to exist at all.

One answer to it is the **Anthropic Principle**. It states that we observe a life-friendly universe because only a life-friendly universe can ever produce anyone to do the observing. That's logically tidy, and it isn't this book's own idea. It's a well-established piece of reasoning in physics and philosophy alike. But it invites a deeper question of its own. Is our universe's stability just a coin toss we happened to win, or does it hint at some larger architecture that favours stable systems from the outset?

Here's a pattern-driven suggestion, and I want to be upfront that this part is mine, a conjecture rather than a settled result.

In most of these trees, matter never binds together, energy disperses without anything useful happening, and no long-lived structure ever gets the chance to form. Most trees sprout and wither almost immediately. But some will be stable enough to support real complexity: stars, chemistry, eventually life. Think about it in family-tree terms. A tree that dies young grows very few branches and leaves few or no descendants behind it. A stable, infinite tree keeps fruiting and branching forever.

Run that forward and the asymmetry becomes enormous. A tree that fails after one generation might produce a handful of branches before it stops. A tree that never fails keeps producing branches without limit, for as long as anything exists to keep counting. Compare those two totals and the short-lived trees barely register. It's not that they don't count. It's that, added up across the whole forest, they never had a real chance of outnumbering the one tree that simply never stopped.

The intuition behind that follows on quite naturally once you sit with it. If some Alt-Multi-Verses are infinite and fertile while others are finite and barren, then the population of observer experiences across everything that exists will be overwhelmingly dominated by the infinite, fertile ones. The Anthropic Principle tells you that you're here because this particular universe allows you to be. My conjecture adds one layer on top of that: you're probably in a stable, effectively infinite Alt-Multi-Verse, because systems like that generate vastly more people like you than any of the alternatives could.

It helps to picture an enormous, unevenly stocked raffle drum. Barren Alt-Multi-Verses each drop in a handful of tickets before falling silent forever. A single fertile, infinite one keeps dropping tickets in without ever stopping. Reach into that drum at random and you'll almost certainly pull out a ticket from the one that never stopped filling.

That's not a claim about being chosen. It's a claim about where the tickets end up piling.

So you can breathe easy as we look a little more closely at what these Alt-Multi-Verses might actually look like.

Some would be close cousins of ours. A universe where gravity runs slightly stronger, or the weak nuclear force runs slightly weaker. Life might still evolve there, differently, perhaps unrecognisably so, but it would still be there.

Others would be almost functional, just off-kilter enough to keep meaningful complexity from ever forming. Imagine an Alt-Multi-Verse nearly identical to ours, except that ice sinks instead of floating. Once the surface of an ocean froze over, that ice would sink, and the layer beneath it would freeze and sink in turn, over and over, until the whole ocean turned solid from the bottom up, taking everything living in it down too. One flipped property, and an entire category of life never gets a foothold.

And still others would be completely nonsensical, laws behaving unpredictably or contradicting themselves from one moment to the next.

A subtler example would be an Alt-Multi-Verse where the strong nuclear force sits just far enough below ours that atomic nuclei never hold together beyond hydrogen. No carbon, no oxygen, no chemistry worth the name. Stars might still burn there, faithfully and forever, but nothing would ever crawl out of them to notice.

The honest answer is that we have no real idea what most of these systems would look like. But we can make one safe assumption. The further a rulebook drifts from ours, the lower the odds of long-lived structure and observers, at least of the kind we'd recognise as such.

Which brings us back to the tree. The Multi-Verse is a single, endlessly branching tree. The Alt-Multi-Verse is a forest of such trees, and a great many of them will never grow very tall. They sprout, twist, collapse and vanish.

We must be living in a section of the forest that grew strong and vast, and that looks likely to keep growing, sustaining each universe and Counter-Verse and Reverse-Verse and every Somewhen and Somewhere that sprouts from them, for as long as the forest holds.

So why are we lucky enough to be here at all?

If our Alt-Multi-Verse really is stable and infinite, then our existence isn't merely possible. In a certain sense it becomes inevitable. There were always going to be observers here. In this exact iteration of this exact universe, you were always going to exist. So was I. And here we both are.

It's possible that conscious life can only emerge under very narrow conditions, in a narrow slice of possible Alt-Multi-Verse parameters. Perhaps the conditions needed for thought, reflection and identity are extremely specific. We're part of a vast forest, but we might be living on one of its luckiest branches.

Some go further still and wonder whether consciousness could exist in entirely different forms in other dimensions, perhaps even forms capable of moving through the forest itself rather than being confined to one tree. That isn't something I'd stake much on personally, but the wider point stands regardless: in something this vast, even the seemingly impossible can turn out to be possible somewhere.

None of this tells us the forest is friendly. It only tells us that if we're standing in it at all, we're very probably standing on one of the branches built to hold weight. That's a claim about where you'd expect to find an observer, not a claim about being watched over, and I'd rather you held it as the former.

It's a strange thing to sit with, this idea that the very fact of your asking the question is itself a kind of evidence. You didn't survive some cosmic filter through luck alone. You're reading this because you're standing on ground that was always going to hold someone. That doesn't make your life any less yours, and it doesn't hand you a purpose you didn't already have. It just quietly removes one more reason to feel like an accident nobody was expecting.

Which is exactly why it's high time we climbed back down to our own tree, our own branch, and looked more closely at the extraordinary things happening right here in this one.

8. BEFORE THE BANG

We've travelled a long way into the mechanics of time, space and the structures that might cradle our existence. There's still an elephant in the room, though. An entire cosmos of questions is pressed into a single moment, and that moment is the one before the Big Bang.

Was there a moment before the Big Bang at all? Did the universe truly emerge from a singularity, a single point of infinite density and temperature? Or was something already there, a field, a pressure, some process already under way?

This chapter looks at a shift in how we understand cosmic origins. It trades the idea of a sharp beginning for something stranger, smoother and possibly far older: **cosmic inflation**.

The story arrives through a chain of insights spread across decades, and it's worth walking the chain instead of jumping to the end.

In the 1920s, the Belgian priest and physicist Georges Lemaître proposed what he called the primeval atom, sometimes the cosmic egg. A compressed origin point expanded outward and gave rise to space and time. Lemaître drew on Einstein's general relativity and on the expanding solutions worked out by Alexander Friedmann, who had shown that the universe's curvature would evolve according to what it contained.

Edwin Hubble then built on the redshift measurements of Vesto Slipher and the distance work pioneered by Henrietta Leavitt. Leavitt's discovery deserves more attention than it usually gets. She found that a certain class of pulsing star keeps time in a way that reveals its true brightness, which gives you its distance. Without that, Hubble would have had a sky full of smudges and no way to say how far off any of them were. Together the measurements showed galaxies moving away from one another, and that meant the universe was expanding.

The Hot Big Bang model that followed explained a great deal. It didn't explain everything.

Look at two opposite edges of the sky. The light arriving from one side has been travelling for nearly the whole age of the universe, and so has the light from the other. Those two regions have never been in contact. Nothing could have passed between them, because there hasn't been time. And yet they sit at almost exactly the same temperature, matching to within a few parts in a hundred thousand. How did two places that never met agree so precisely on anything?

Then there's the shape of space. Space appears to be almost perfectly balanced, geometrically speaking, and that balance is unstable. Any tiny deviation early on would have been amplified enormously over billions of years, tipping the universe into collapse or into runaway dilution. For it to look balanced now, it must have started out balanced to an absurd degree of precision. That isn't impossible, but it's the kind of coincidence that ought to make you suspicious.

In 1980 the physicist Alan Guth proposed an answer to both problems at once. Before the classical Big Bang, he suggested, the universe went through a brief phase of exponential expansion. In a fraction of a second, space ballooned by a factor of at least 10^26 in size.

Sit with that number for a moment. It means something the width of a proton stretched out to something roughly the span of a galaxy, and it happened in far less time than it takes light to cross an atom.

Both problems dissolve if that's what occurred. The distant regions of sky do match because they were once close enough to reach thermal agreement, before inflation flung them apart. And space looks balanced because stretching anything by that factor flattens it, the same way a wrinkled sheet pulled taut across an enormous frame ends up smooth wherever you happen to be standing on it.

Andrei Linde extended Guth's work with what he called chaotic inflation. In his version, inflation might not be a one-off event at all. It could continue endlessly in the wider structure, with regions dropping out of it here and there to become universes of their own, while inflation carries on elsewhere forever.

That should sound familiar. It's the Multi-Verse from Chapter 6 arriving by a completely different road. We got there before by asking what happens to unrealised quantum outcomes. Linde gets there by asking what inflation does if you simply let it run. Two unrelated lines of reasoning, both ending in a great many universes, and neither one leaning on the other for support. When that happens in physics it doesn't prove anything, but it does tend to make people look harder.

Worth noting too that these bubbles need not share our constants. Regions that stop inflating at different moments, or in different ways, could settle into different physics entirely, which puts Linde's picture closer to the sixth rung of our ladder than the fifth.

The astrophysicist Ethan Siegel has done a lot of the work of explaining what this actually implies, and his framing is the one I find clearest. Inflation isn't a patch applied to the Big Bang. In many models it comes first and creates the conditions for it. On that view the hot Big Bang isn't the beginning of everything. It's the end of inflation, a reheating event where the energy driving the expansion decays into matter and radiation.

That's a real reversal of the usual picture, and it's worth stating plainly. The Big Bang stops being the opening scene. It becomes the moment the lights came up on a stage that was already there.

Reheating is worth a moment on its own, because the name is doing a lot of quiet work. During inflation the universe is close to empty of ordinary matter, and bitterly cold. All the energy is locked up in the field driving the expansion. When that expansion ends, the field dumps its energy into particles, and the temperature leaps from almost nothing to something like a billion billion billion degrees. That surge is the fireball we've always called the Big Bang. It isn't an explosion in the ordinary sense. It's a phase change, closer to steam condensing than to a bomb going off.

Inflation makes a sharp prediction about the shape of space. On large scales the universe should have no overall curvature. Triangles should add to 180 degrees, and parallel lines should stay parallel however far you follow them.

Physicists call this a flat universe, which is unfortunate, because flat suggests a pancake and this has nothing to do with being two-dimensional. What's meant is straight geometry in three dimensions. To avoid the confusion I'll call it a **Straight Universe** throughout this book.

In a Straight Universe, light travels in straight lines across cosmic distances. Large-scale curvature sits at zero. Expansion may well continue forever, and the geometry stays balanced while it does.

The observational evidence for this is about as strong as cosmology gets. To see why, we go back to the baby picture of the universe: the Cosmic Microwave Background, the faint afterglow released around 380,000 years after the Big Bang, when the universe finally cooled enough for light to travel freely.

Three missions sharpened that picture in turn. COBE, running from 1989 to 1993, first detected the tiny temperature variations in the afterglow. WMAP, from 2001 to 2010, mapped those variations far more precisely. Planck, from 2009 to 2013, produced the most detailed map of the early universe we have.

All three confirmed predictions inflation had made in advance, which is the part that matters. A theory that explains what we already knew is cheap. A theory that tells you what you'll find before you look is doing real work.

The afterglow turned out to be almost uniform, with only faint variations, exactly as you'd expect if inflation smoothed things out. The variations came in a nearly scale-invariant spectrum, meaning they have similar strength across a wide range of sizes, with a slight tilt in a particular direction. Inflation predicts that tilt. And there are correlations across patches of sky larger than could have been in contact at the time, which is what you'd get if inflation had stretched quantum ripples out across the whole sky.

Those patterns fit inflation. They don't fit the original Hot Big Bang model without it.

So why care about the geometry of the universe at all? Historically, geometry and fate were thought to be locked together.

The reasoning went like this. A closed, positively curved universe would eventually collapse in a Big Crunch. An open universe would expand forever. A straight universe would balance on the edge between them, expanding forever but always slowing down.

Given that we appear to live in a Straight Universe, we should therefore expect the expansion to be gradually slowing.

Then in the late 1990s, observations of distant supernovae showed that expansion is accelerating instead. Whatever drives that acceleration acquired the name dark energy, which is a label for our ignorance more than an explanation. If it stays roughly constant, then even a Straight Universe expands forever, and does so faster and faster.

Not every physicist is convinced dark energy is constant, and the jury remains out. There's room here for other accounts, including speculative ones like the Entropic Tension Force we looked at in Chapter 5.

Now I owe you something from Chapter 3.

When we looked at the early universe, I said we'd come back to the question of what caused the quantum fluctuations that seeded every structure we see. Inflation supplies the answer, and it's my favourite idea in cosmology.

Quantum fields are never perfectly still. Even in what we'd call empty space, they jitter, with tiny fluctuations appearing and vanishing constantly. At normal scales this is utterly negligible. During inflation it wasn't, because inflation took whatever was there and stretched it.

Fluctuations that began at subatomic scale were pulled out to astronomical scale in a fraction of a second, and then the expansion outran the ability of those regions to smooth themselves back out. The jitter froze in. It stopped being quantum noise and became a permanent map of slightly denser and slightly emptier patches of space.

Gravity did the rest, over billions of years, pulling matter toward the denser patches until they became galaxies.

So every galaxy, every star, every planet, and every person traces back to a quantum fluctuation that happened to get caught mid-flicker and stretched to the size of the sky. The largest structures in existence are the fossilised imprint of the smallest possible events. I find that one of the more astonishing claims modern physics makes, and unlike much of what this book discusses, it has serious observational backing.

That's the standard account and it does its job well. I want to put one more thought beside it, clearly labelled as mine and as speculation.

In Chapter 5 we imagined two runners tethered by elastic, and the tether only makes sense if both ends are doing something. If ETF is real, then it should show up at both ends of a universe's story rather than only one.

Consider the clustering first. In a Counter-Verse pairing, the moment just after our Big Bang would correspond to the moment just before our partner's ending. At that end, entropy is enormous. Galaxies are collapsing, black holes are merging, and density contrasts are extreme. If entropy really is shared across a boundary instead of being sealed on one side of it, then the pattern of extremes over there might counter-point onto the pattern of faint irregularities over here. Where they have a great concentration, we might have a corresponding feature in the seeds of ours.

This wouldn't replace quantum fluctuations. It would sit faintly on top of them, and the honest position is that we've no way at present to tell the two apart.

Now consider the acceleration. In a Reverse-Verse pairing, where the same universe runs back the other way, you get something more like a slingshot than a counter-point. Entropy climbs toward the boundary, and instead of stopping there it carries through and reverses direction. From inside, during the approach, that would look like an expansion that isn't settling down the way the geometry says it ought to.

I'd stress again that this isn't testable in any form I can offer you, and that dark energy has perfectly respectable conventional explanations. It's an idea about how the pieces might connect if the earlier chapters happened to be pointing at something real.

There's a harder tension I promised in Chapter 4 that I'd bring back here instead of letting it quietly drop.

The CPT-symmetric model doesn't need inflation. Its authors treat that as a strength, since it accounts for the smoothness of the early universe without an extra epoch of exponential expansion. This chapter has just made a fairly enthusiastic case for inflation on the strength of predictions that were confirmed after they were made.

Both of those can't be the last word. I'm not going to pretend otherwise, and I'm not going to shelve one to protect the other. Inflation has the better observational record at present. The CPT model has an elegance and an economy that inflation doesn't. If a decisive test arrives, one of them goes on the shelf, and that's how it should work.

An idea-system can hold two accounts that don't reconcile, provided you keep the disagreement in view instead of papering it over. That's the whole reason for building one.

Back to inflation, then, and to the obvious next question.

If inflation preceded the Big Bang, the universe before the hot plasma was likely dominated by a scalar field, usually called the **inflaton**. The potential energy stored in that field drove the rapid expansion. When inflation ended, that energy decayed into matter and radiation, and that decay is the hot Big Bang we've been calling the beginning all along.

But where did the inflaton come from, and where did its energy come from?

One possibility comes from Loop Quantum Cosmology, which we met in Chapter 4. At ultra-high densities, in that framework, gravity becomes repulsive instead of attractive. A contracting universe doesn't collapse into a singularity. It reaches a floor, bounces, and expands again.

That bounce might supply exactly the energy conditions inflation needs. Inflation stops being the beginning and becomes the bridge.

At those temperatures and densities, the distinctions we rely on start to break down anyway. Space, time, energy and matter stop being clearly separable things. The inflationary field may not sit inside space at all. It may be space, or whatever space was before it became the sort of thing we can stand in.

However we cut it, the shift is this. Inflation may not patch the Big Bang. It may produce it. The sharp beginning we used to picture gets replaced by a smoother ramp that made the hot universe possible in the first place.

So let's take the wider view.

The further we go, the more beginnings and endings look like artefacts of scale. Time may not begin or end at all. It may expand, contract, reverse or bounce. The universe may not be a single thing, but part of a network in which entropy, energy and information are conserved and transformed rather than created and destroyed.

There's a strange comfort in that.

If the story of the universe didn't begin at the Big Bang, then perhaps it doesn't really end either. Perhaps we're part of a cycle. A breath in a longer rhythm. A ripple in something considerably larger.

Maybe what matters most isn't whether the universe had a beginning. Maybe it's that the universe has continuity, and that we belong to something vast, lawful and still unfolding.

9. LAWS BEYOND LAWS

We've come a long way. We started on familiar ground in our own universe, moved out through the branches of the Multi-Verse, and then out again to the Alt-Multi-Verses, where the laws of physics themselves begin to shift.

That leaves us with a question we can't dodge any longer. If each Alt-Multi-Verse runs its own rules, its own gravity, its own quantum limits, its own thermodynamics, then what holds all of them? What is the **Omni-Verse** actually made of?

Let me start with something small enough to picture.

Take a jar down off a shelf. Fill it with water, seal it, and put it over a flame. Inside the jar, things start behaving very differently from the air in the room. Water turns to steam, pressure builds, and if you keep going the whole thing will eventually give way.

But the jar isn't outside physics. It sits in the room, and it obeys the same gravity, the same thermodynamics and the same passage of time as everything else in the room. The jar has its own local conditions. It doesn't have its own laws.

So here's the question this chapter turns on. What happens when the jar is big enough to hold whole Alt-Multi-Verses? Does that jar need to sit in a lawless void? Or is there a shelf it rests on, some higher framework, a law of laws?

We're at the outer edge of speculation now, and I'll say so as we go. But these aren't idle ideas. Serious physicists and philosophers have spent careers on them.

Let's start with what string theory does to the question.

In string theory, particles aren't point-like dots. They're tiny vibrating strings, and how a string vibrates determines what kind of particle it is. What governs the vibration is the shape of the universe's extra dimensions, the ones curled up too small for us to notice.

Here's the trouble. There are an enormous number of ways those dimensions could be folded. The figure you'll see quoted is around 10^500 possible configurations, and it's worth knowing where that number comes from. It isn't a derived result. It's an order-of-magnitude estimate from counting one particular family of ways to wrap the extra dimensions, and different counting methods give different answers. The precise figure matters less than the scale of it. No ordinary word covers a number like that.

Each configuration yields a different set of physical laws, which is to say a different Alt-Multi-Verse.

This is known as the string theory landscape problem, and it's a problem instead of a triumph for a specific reason. The theory turns out to be so flexible that it permits almost any universe you like. It doesn't tell us why we ended up in this one. It doesn't tell us how to choose between them. It doesn't even tell us whether the others are real or merely permitted on paper.

A theory that can accommodate any observation isn't obviously telling you much about the world. That's the honest criticism of the landscape, and plenty of physicists make it.

In jar terms, string theory hands us a shelf with an unimaginable number of jars on it, and no account of why we're in this one.

Which brings us to a bolder answer.

The physicist Max Tegmark proposed what he calls the **Mathematical Universe Hypothesis**. His claim is that every mathematically self-consistent structure is real in some form. If the equations hold together, the universe they describe exists. Not "could exist" or "might have existed". Exists.

It's an elegant move, because it dissolves the selection problem entirely. There's no need to explain why our laws were chosen, since nothing was chosen. Every consistent set is out there, and we're in one of the ones that permits people.

But notice where it leaves the jar. Tegmark's answer implies a kind of library holding every self-consistent structure, and that library isn't obviously governed by anything. So we're back where we started, asking what contains the container.

There's a further oddity worth sitting with. If mathematical consistency is all it takes, then the vast majority of existing structures would be nothing like a universe at all. Most self-consistent mathematics describes no space, no time and no change. It just sits there, true and inert. Worlds with anything happening in them would be a vanishingly small corner of everything real, and we'd be in that corner for the same reason we're on a fertile branch in Chapter 7. Nowhere else can hold an observer.

Tegmark's shelf, then, is less a shelf of jars than a shelf where almost everything is solid glass, with a few hollow ones scattered through it.

A different approach treats the whole thing as computation.

Edward Fredkin and Stephen Wolfram, among others, have argued that the universe behaves like an unfolding algorithm running on digital rules. In **digital physics**, the universe is information first and matter second. Perhaps it's something close to code.

This connects to the idea of meta-laws. These aren't rules for how our universe behaves. They're rules for how universes get generated at all. If every Alt-Multi-Verse is something like a running process in a larger computational framework, then the framework might need only one rule: each process must be internally consistent.

Picture a filing system. Each folder holds one Alt-Multi-Verse with its own rulebook and starting conditions. Inside that folder sit countless subfolders holding the variations across dimensions five through nine. The operating system can open and run any of them, so long as they obey basic logic. That operating system is the meta-law, and the jar analogy shifts. We're no longer asking what the shelf is standing on. We're asking what the shelf is written in.

It's a tempting picture, though I'd flag what it quietly assumes. Every generation reaches for its most advanced machine when it wants to describe the universe. Newton's contemporaries saw clockwork. The Victorians saw steam engines and heat. We see computation, and the fit feels perfect. That doesn't make it wrong, and information really does seem to be doing something fundamental in modern physics. It's worth keeping one eye on the possibility that we're describing our tools instead of the world.

From there it's a short step to a theory most people have already heard of.

Simulation Theory asks whether our reality might be a simulation created by some advanced intelligence. The philosopher Nick Bostrom gave the argument its best-known form in 2003, and it's usually reported more crudely than he actually put it.

Bostrom's argument isn't a flat claim that we're probably simulated. It's a trilemma. He argues that at least one of three things must be true. Either civilisations almost always go extinct before reaching the technology to run detailed simulations of conscious beings. Or civilisations that reach that stage almost never bother to run such simulations. Or we're almost certainly living in one.

Put that way it's a sharper argument, because the conclusion is a fork instead of a verdict, and each branch tells you something uncomfortable.

I'd add a caution about the third branch. The reasoning compares a vast number of simulated realities against a single base reality, and that comparison does a lot of the work. It also assumes a great many civilisations have the technology, the resources and the inclination to run enormous numbers of conscious simulations. That's three separate assumptions dressed as one.

Change the denominator and the maths changes with it. If there are infinitely many base realities, as the branching pictures in earlier chapters would suggest, the ratio stops being so lopsided. That doesn't rule the idea out. It does mean the confident version of it rests on a narrower base than it appears to.

What would it mean for physics if it were true?

Our laws would be constraints written into the program. Our constants, forces and dimensions would be settings specific to this build of the software. And the engine running it might be running many others alongside, each with its own settings, which puts us back at Alt-Multi-Verses by another route. Just as a video game runs on design rules its players never see, whole universes might be governed by constraints invisible from inside them.

The philosophical answers go somewhere different, and they're worth taking as seriously as the physical ones.

Idealism holds that reality is fundamentally mental rather than material. On this view the physical world is a manifestation of consciousness or thought instead of the other way around. George Berkeley argued in the eighteenth century that existence depends on perception. More recently the cognitive scientist Donald Hoffman has argued that what fundamentally exists is a network of conscious agents, and that the physical world we perceive is closer to an interface than to the underlying reality, in much the way a desktop icon tells you nothing about the circuitry beneath it.

Panpsychism takes a related but distinct line. It holds that consciousness is a fundamental and widespread feature of the universe, present in some minimal form throughout it. Versions of the view have been attributed to William James, Alfred North Whitehead and Bertrand Russell. The philosopher Philip Goff has argued for a modern form of it, suggesting that all physical entities possess some rudimentary form of experience.

These aren't ideas I hold. But I can see clearly how each becomes load-bearing for someone building an idea-system, and that seems worth understanding instead of waving away. Each one is a serious attempt to solve a real problem. Nobody has managed to explain how physical processes give rise to experience at all. Dismissing that problem is easier than answering it.

Then there's the most extreme answer of the lot.

Modal Realism was championed by the philosopher David Lewis, and it claims that all possible worlds are equally real. Not only worlds permitted by mathematics or generated by simulations, but any coherent possibility whatsoever. If a reality can be described without contradiction, it exists. It doesn't need to be built, discovered or run. It simply is.

That's the jar analogy taken to its conclusion, and the conclusion is that there is no jar. No shelf, no room, no law of laws. Just an endless sea of realities, in which even the absence of familiar logic might hold somewhere.

Lewis was serious about this in a way people often assume he couldn't have been. He didn't mean these worlds exist as ideas or as possibilities we entertain. He meant they exist exactly as concretely as this one, with people in them as real as you, who would call their world actual and ours merely possible. The word "actual" on his account works like the word "here". It doesn't mark out anywhere special. It just points at wherever the speaker happens to be standing.

Which raises an obvious objection, and it's one worth answering carefully.

Does that mean every world we can imagine exists?

The short answer is no. Imaginable and possible aren't the same thing. Possible, in the sense every theory in this chapter uses, means internally consistent. It means a world that doesn't quietly break its own rules.

One of my small frustrations, as someone who loves science fiction and fantasy, is a story that tears up the rulebook it wrote. I'll suspend disbelief happily. But if you establish in the first chapter that a character is nearly invulnerable, you can't have them yelp at a thumbtack in the second, unless you've also given us the exception, some weakness or cost or spell that accounts for it.

Take Ant-Man. If your rulebook says mass and gravity work as they do here, then shrinking a person to the size of an ant while keeping their mass creates immediate trouble. His volume drops by a factor of tens of millions, so his density goes through the roof. Surface gravity at his skin would soar. His inertia would make a thrown punch land like a micrometeorite. No flying insect could lift him. Yet he lands lightly, rides an ant, and throws people around without punching through the floor.

That's a rule conflict, and it's the sort that keeps a world out of the library.

You could fix it. Have the suit shunt most of his mass into a reservoir somewhere else, another dimension or field, and retrieve it when he grows. That's a consistent rule, and it even generates interesting consequences of its own. But unless the fix holds across every character and every scene, that world can't exist in the exact form it's presented.

Fiction bends its own rules for drama, and that's entirely fine for art. In the landscape we're discussing here, though, possible means non-contradictory. A world can be as wild as you like. It still needs a rulebook it doesn't secretly tear up when the plot gets inconvenient.

So when these theories say "coherent without contradiction", that's what they mean. No hidden exceptions smuggled in after the fact. No outcomes that violate the very laws the world claims to run on. Only those worlds earn a place on the shelf.

This is a more demanding filter than it first appears, and it does real work in trimming the sea of possibilities down to something smaller. A world can have magic in it, provided the magic obeys a rulebook that holds every single time it appears. A world can run time backwards, or have gravity push instead of pull, or dispense with matter entirely, so long as it doesn't help itself to a convenient exception the moment the consequences become awkward. What gets excluded from the library isn't strangeness. It's incoherence, and that turns out to be a different thing altogether, and a much rarer one than most people assume when they first hear that everything possible exists somewhere.

Contemplating this territory may be more useful as philosophy than as physics, and I'd not pretend otherwise. It still does something valuable. It shows how hard it is to draw a boundary around the phrase "what exists".

Each of these views is a different language for the same mystery. Whether any of them turns out to be true matters less, for our purposes, than what they reveal about how we think about existence. As with everything in this book, the aim isn't belief in one. It's a fair look at all of them.

So what does this leave us with?

As we push outward we eventually hit the limit of what experiment can reach, and we cross into metaphysics and interpretation. That boundary is real and I'd not blur it. But these ideas still matter, because they shape how we hold our own reality.

If we live in a simulation, or inside a mathematical structure, or on one branch of an endless law-generating forest, then our sense of what's fundamental has to shift.

It opens a paradox too. The deeper we go, the less we can assume about the rules. At some level even the idea of a law may stop making sense. The laws of our universe may not be universal at all. They may be local settings, true here and nowhere in particular else.

Here's the comfort in that, and I think it's real.

Even if we live in a simulation, or a contained structure, or a jar on somebody's shelf, our experience isn't diminished by it. Meaning doesn't leak out through the boundaries of a system. If anything, it takes its shape from them.

The code still runs. The jar still holds water. The forest still grows.

And when we keep asking questions across dimensions and disciplines, we're doing something worth doing. We're taking part in the unfolding of curiosity, inside a reality that permits curiosity to exist at all.

10. THE SHAPE OF BELIEF

Through scientific discovery, reason and sheer persistence, it's becoming possible to evaluate and sometimes even validate theories about how the universe came to be, how it works, and what that might mean for our lives. Humanity has of course been at this since the beginning. We like to think we're cleverer than our ancestors. More often we're simply luckier, standing on centuries of preserved knowledge, formal education and cumulative error-correction. If progress keeps unfolding, a good deal of what we treat as bedrock will look quaint to future minds. The best any era can manage is to build on what it inherited, accepting that insight and error travel together from one generation to the next.

We benefit from a long relay of discovery, stacked up in books, code, data and algorithms. Who can say what Galileo or Hypatia might have worked out with access to quantum papers and images from Hubble and the James Webb telescope? As we audit today's models, it's both humbling and clarifying to notice how many ancient worldviews, often dismissed as primitive, were aiming at the questions we now approach through science.

The difference between them and us is mostly one of instruments and accumulated corrections, not of raw intelligence. A person born four thousand years ago had the same brain you have. What they lacked was three millennia of other people writing down where they went wrong. That's not a small thing, and it's the entire reason we can do what we do. It's also worth remembering when we're tempted to feel superior about it.

It's vital that we learn to examine conventional thinking critically. It's equally foolish to ignore scientific consensus when nothing robust contradicts it. Look at flat-earth arguments, or anything of that shape, and you'll usually find they depend on the critical thinker setting aside the overwhelming majority of the observational evidence. That's not scepticism. It's selective attention wearing scepticism's coat.

Ultimately, if someone has decided they want to believe something, there's very little anyone can say to shift it. That's the trap of entrenched belief, whether it's religious, political, or scientific belief that has hardened into ideology.

And yet it's striking how often early cosmologies arrived at metaphors that echo themes in this book and in modern physics. Cycles of time, sacred dualities, hidden forces, even the idea that reality itself might be a kind of illusion. They reached these independently, across continents and millennia. These are not new ideas.

One of the oldest surviving cosmologies comes from the Indian subcontinent, from around 1500

BCE. Hindu cosmology describes a universe that expands and contracts through an infinite series of cycles, each lasting billions of years, in a process not far off modern cyclic or bouncing cosmologies. Creation, preservation and destruction, associated with Brahma, Vishnu and Shiva, run parallel to expansion, stability and collapse. The yugas, or ages of the world, describe vast epochs with differing qualities and differing degrees of order. Our current age, Kali Yuga, is said to be marked by moral and cosmic decline, which chimes with scientific conversations about the long-term fate of the universe and the steady climb of entropy.

There's a second idea in that tradition worth pausing on, because it turns out to be the sharpest of the lot. Hindu and Buddhist thought both carry the concept of **maya**, usually translated as illusion, though that translation flattens it. Maya isn't the claim that nothing exists. It's the claim that the solid, observable world we move through is a kind of perceptual shell, and that whatever is fundamental sits beneath it, less tangible and far less obliging to our senses.

Set that beside what physics has spent the last century saying. Solid matter turns out to be overwhelmingly empty space. The apparently fixed objects around you are localised excitations in fields that extend everywhere. Time doesn't run at one universal rate, and simultaneity depends on who's asking. The world your senses report is a workable summary built for staying alive, and it isn't the world as it is.

I'd not push the parallel too hard. Ancient thinkers weren't doing quantum field theory in disguise, and any resemblance is a matter of pattern instead of derivation. But of all the metaphors humanity has produced for the gap between appearance and reality, this one has aged remarkably well.

From around 2600 BCE, the Egyptians developed a worldview governed by **Ma'at**, a principle of truth, balance and cosmic order. Life, death and rebirth weren't endings but transitions inside a larger harmony. In death, a soul was weighed against a feather. Light enough and it could ascend. Too heavy and it was devoured.

That isn't science. It is, though, a moral encoding of balance, something close to a metaphysical conservation law. Much as thermodynamics does in modern physics, Ma'at functioned as the thing holding equilibrium in place across the whole cosmos.

By 400 BCE, Chinese Daoism had formed a worldview grounded in flow and harmony instead of rigid structure. The **Dao**, meaning the Way, is the unseen force behind all things, and existence takes its shape from the interplay of yin and yang. Those aren't opposing forces so much as complementary ones, each requiring the other to mean anything at all.

That sits curiously well beside quantum field theory, where particles aren't fixed objects but localised excitations in fields, always in flux. Daoism's emphasis on natural balance also anticipates the systems thinking we now use in ecology and physics, where interdependence instead of dominance is what keeps a system going.

The Greek Stoics, around 300 BCE, introduced the **Logos**, a rational ordering force structuring the cosmos. The Stoics held that the universe was material, deterministic and governed by reason. Time for them wasn't linear but cyclical, and even the destruction of the universe, which they called ekpyrosis, would be followed by its rebirth. That anticipates the scientific determinism running from Newton to Laplace, and it speaks to an interconnectedness of all things that some readings of quantum entanglement still gesture at today.

In Mesoamerica, between 200 and 1500 CE, cultures including the Maya and the Aztecs held deeply cyclical views of time. The Mayan Long Count calendar tracked immense cosmic cycles of creation and destruction. Their calendars weren't simply for keeping dates. They tied human action to cosmic rhythm, and they treated the heavens as alive, active and bound up with the fate of human beings. Once again we find a model that mirrors today's cyclic universe theories, particularly the ones proposing collapse and rebirth on a universal scale.

In Norse cosmology, between 800 and 1100 CE, the universe was structured across multiple realms connected by **Yggdrasil**, the World Tree. Time ran in a line, but that line was destined to end in Ragnarök, a great undoing of the cosmic order followed by renewal. The gods fall, the world is destroyed, and the world rises again.

The parallels here run in two directions. Ragnarök resembles Big Crunch and heat death models, with the significant addition of a reset. And Yggdrasil itself connects realms in a way not unlike the brane cosmology we met in Chapter 6, where universes sit layered in higher-dimensional space, near each other and yet unreachable from within. A tree whose branches hold separate worlds is a surprisingly serviceable image for that, and the Norse got there without the mathematics.

Celtic and other European pagan systems often centred on spirals, knots and seasonal cycles, all symbols of interconnectedness and eternal return. The **Ouroboros**, the serpent swallowing its own tail, is among the most enduring of them. It represents continuity, self-reflection, and the cycle of death and rebirth. It isn't hard to see the resonance between that and modern physics' interest in closed-loop systems, conservation laws, and the possibility that time is more circular than we tend to assume.

So why do so many cultures arrive at the same handful of ideas? The same themes keep surfacing: cycles, balance, flow, decline and rebirth. Is this divine inspiration, or is there a more obvious explanation available?

I think the answer is simple and rather satisfying. We all observe what's visible, and we use it to imagine what isn't.

That habit is deeply adaptive. If you tell me sabretooths roam a particular valley, and I hear something roar in that valley, I don't wait to verify tooth morphology. I run. Evolution rewards the educated guess far more reliably than it rewards the careful one. So we watch tides and seasons, and then we scale those patterns up to stars and galaxies, because those are the patterns we have.

Recurring forms in nature produce recurring metaphors in culture. And the forms really do recur. The same force that pulls an apple down beside Isaac Newton pulls a passing rock into a black hole. It's natural that we reached for harvests, seasons, ageing and renewal to make sense of something as immense as the universe. It's equally natural that some structures really do repeat across scales, so that the metaphors occasionally land closer to the mark than their authors could have known.

Similar themes don't make these systems accurate, though, and they couldn't all be accurate at once. Where ancient models diverge from science isn't usually in the pattern they noticed. It's in the motive they projected onto it.

Why is it raining? Because a god is displeased. The Sun isn't merely rising, it's being dragged across the sky by a charioteer. Harvests don't fail because of weather, they fail because we failed someone. Anthropomorphism hands the universe an agency it probably doesn't have, and that move does two things at once. It supplies meaning, which is a real human need. It also justifies power, and sometimes violence.

People have long explained their own survival by supposing they were chosen. Cosmology wasn't only about how the stars moved. It was used to explain why the harvest failed, why the plague came, and what a people were destined for. Belief systems arose to explain, certainly, but also to reassure and to justify.

Consider how well that works as a machine. If a people want a neighbour's land, and they believe the gods chose them specifically, the invasion arrives pre-justified. If the invasion succeeds, it's their children who prosper and carry the belief forward. The belief system doesn't merely survive. It thrives, precisely because it was useful to the people who won. Nothing about that process filters for truth.

And as belief systems settle, they tend to harden. A flexible metaphor like "the Sun is pulled across the sky" becomes a fixed truth: Apollo rides his chariot each day, and questioning it is heresy. Once belief ossifies it stops exploring and starts policing.

Modern minds are not exempt. Ask a committed atheist what colour Santa's suit is and they'll tell you confidently that it's red, which it is, and has been in illustration for well over a century. Ask them what colour to paint a baby boy's room and the odds are good they'll say blue. Yet pink was once considered the boy's colour, on the reasoning that it sat closer to red and red was the colour of blood and vigour. Western culture reversed the convention within living memory, and the new arrangement now feels so obvious that most people never think to ask where it came from.

We may believe we've moved beyond rigid ideology. Mostly we've just updated our myths and stopped noticing that they're myths.

Science is not immune to the same hardening, and it would be dishonest of me to pretend otherwise. Alfred Wegener proposed continental drift in 1912 and was dismissed for roughly fifty years by people with excellent credentials, long enough that he died before the seafloor evidence arrived to vindicate him. Barry Marshall had a shorter wait and a stranger method. Unable to persuade anyone that stomach ulcers were caused by bacteria rather than by stress and stomach acid, and unable to find an animal to test it on, he drank a beaker of broth swimming with the bacteria himself in 1984. He had gastritis within a week. He and the pathologist Robin Warren took the Nobel Prize for it in 2005. The difference isn't that scientists are free of the tendency. The difference is that the method eventually forces the correction through, whether the practitioners enjoy it or not. That's the whole value of it. A framework that can be embarrassed by evidence is worth more than one that can't.

Here's the clearest example I know, and it's one nobody thinks of as a belief at all. Money has no intrinsic value. A banknote is a piece of paper, and a number in an account is a pattern in a database. Neither will feed you, shelter you, or keep you warm. Money works only because everyone involved agrees that it works, and that agreement is so complete and so continuously reinforced that it feels like a property of the physical world instead of a shared decision.

It also happens to be one of the most powerful forces shaping human lives. Wars are fought over it, and people organise their entire existence around acquiring it. A collective belief with nothing underneath it is doing all of that, every day, and we barely register it as a belief.

I raise it because it makes the point better than any argument about religion could. Shared belief isn't a primitive habit we grew out of. It's load-bearing infrastructure. The question was never whether to hold beliefs in common with other people, since we obviously must. The question is whether we notice when we're doing it.

We like order. We crave systems. We also tend to confuse consensus with truth, largely because uncertainty is uncomfortable to sit with.

As a Catholic child I was taught not to eat meat on Good Friday. The implication wasn't only that I'd be breaking with tradition. It was that breaking the rule might affect my soul. The tradition is widely thought to have roots in historical fish surpluses and practical economics rather than in anything theological. Even so, it carried far more weight than nutrition ever could. It carried meaning and consequence, and it shaped how I understood myself as a Catholic child.

As it happens I quite like fish, so it isn't a bad memory. For others it may be a cherished tradition, and I've no issue at all with its preservation as part of the culture of Catholic households. What I take issue with is the moment such patterns become tests of virtue, sorting people into the chosen and the less than. From there it can be a very slippery slope.

None of this is an argument against belief systems. They carry meaning, cohesion and care, and they've held communities together through things that would otherwise have scattered them. It's an argument against brittleness, the particular kind that blocks learning.

My aim isn't to adjudicate any creed. It's to show how rigid ideologies, religious or scientific alike, can put a ceiling on personal and intellectual growth. This book doesn't deliver a final answer. It explores a number of plausible frameworks and invites you to handle them. No chapter here is gospel, and no idea in it is unassailable, including my own.

I don't subscribe to the religious systems I've mentioned, but I value their metaphors. Many of them still shine. Sift patiently and you'll find gems, images of balance, return and responsibility, that harmonise surprisingly well with the models we build today.

In the next chapter we'll revisit a few of the scientific frameworks you've already met, and ask what they might mean for a person trying to live well, here and now. Not to hand down conclusions, but to pose better questions and offer some lenses you can try on in your own idea-system.

11. ONE LIFE IN THE OMNI-VERSE

In Chapter 2 we explored Block Theory, sometimes called Eternalism, the idea that past, present and future all exist at once. Time, on this view, isn't a river flowing past you. It's a dimension like space. Just as every place exists whether or not you happen to be standing in it, every moment exists whether or not you happen to be experiencing it. The implication runs deep. Time stops being a process and becomes a structure.

That raises a hard question. Does free will survive a model like that? If we picture time as a single fixed line, with each action leading to exactly one outcome, then it starts to look as though we're destined to think every thought we'll ever think, to make every decision we'll ever make, and to live only the one life already laid out ahead of us.

But in Chapter 6 we met a real counterweight to that. The fifth dimension, the level at which alternate timelines unfold, put choice back on the table.

The fifth dimension isn't another direction through space or time. It's a higher-dimensional reality in which more than one timeline plays out. On this view the present moment isn't the only path forward. Every decision you face becomes a branching point. Front door or back door? In the fifth dimension, both futures exist. Each of those timelines remembers you as its past, and then diverges depending on which door you actually took. You, as a four-dimensional being, will only ever remember one of those futures. Other versions of you remember the road you didn't walk.

You can't travel backwards and change the path you've already taken. The arrow of time still has you. The version of you who trained at the gym every day for five years gets the results of that. The version who ordered takeout every night gets a different set of results. You can't undo either choice. But neither are you limited to just one version of yourself going forward from here.

So do those other versions of you actually exist, or is this just a nice way of talking?

Here's a way into it. As I write this, I'm sitting in Ireland. I don't live in America. Does America still exist? Obviously it does. Not being present somewhere doesn't make that place stop being real. That's object permanence, something you developed as an infant somewhere between four and twelve months old. It's the realisation that things keep existing even when you can't see them. A toy hidden under a blanket is still there. A parent who's left the room is still there, just out of view.

When we talk about **Somewhen**, we're applying that same intuition to time. Call it **Temporal Permanence**. Just as the toy stays real under the blanket, the past and future stay real even when you can't perceive them directly. The Multi-Verse stretches the idea further still. There isn't only one fixed past and one fixed future. There may be a vast ensemble of them, each branch a genuine expression of a choice you might have made, a life you might have lived.

That gives grief a gentler shape than it usually gets. A baby mourns a vanished toy right up until object permanence settles in. We mourn the people we've lost because they feel gone. But if time is block-like, and the Multi-Verse is real, then they aren't gone everywhere. They exist somewhen, in the past of this universe, or in a version of events unfolding differently at this same clock time. We can't visit those other realities. That doesn't stop them being part of a larger truth.

Perhaps in one of them your loved one lived longer. Perhaps the friendship never broke. Perhaps you're the one who died instead. In the full landscape of possibility, every version of love, loss, redemption and regret plays out somewhere. We don't perceive each of those lives. They're lived nonetheless, by parallel versions of us.

Does any of this make this life smaller? I don't think so. If anything it raises the stakes. What you do here becomes your own particular expression of self, and lived fully it's a meaningful thread in something much larger. Far from cheapening life, the Multi-Verse asks for a kind of reverence. Every person you meet is a crossroads of possible histories, and this meeting, the one happening right now, is the one you both actually get to keep.

Somewhere there's a reality where Albert Einstein, with all his potential intact, ended up uneducated and homeless, written off by everyone who passed him on the street. Maybe he shared his ideas anyway, and people laughed them off as nonsense. There's very likely a reality where far greater minds than his have shaped humanity, and in that same reality, children who might have become those minds never got access to clean water and didn't survive childhood. Somewhere in the Multi-Verse, each of us, at some point, in some reality, probably contributes something that actually matters. The Multi-Verse doesn't guarantee greatness. It does suggest the space of possibility is far richer than it looks from inside a single path.

I sometimes wonder what my life would look like if I'd been born a thousand years from now, or a thousand years ago. A different country. A different career. The creative urge I never quite chased seriously enough. If the Multi-Verse is real, those lives exist somewhere. That doesn't make this one smaller. It makes it larger. My unfulfilled dreams are fulfilled somewhere, and plenty of what I did find here, the actual joys of this actual life, outshine most of the alternatives I can picture. I try not to pine for other branches. I try to appreciate this one, and to act in ways my future self will thank me for.

Across the full range of possibility there are versions of you who are wiser, more patient, crueller, more generous, more frightened, braver. Some loved more completely. Some betrayed. Some never got the chances you got. Some got luckier than you. Does any of that make what you do here matter less, cancelled out by some better version of you elsewhere? No. Meaning doesn't get imported from other branches. It gets made in this one.

Kant argued that moral worth lives in the will behind an act rather than in its outcome. Act only in a way you could wish, without flinching, that everyone else would act too, was roughly his standard. In an Omni-Verse that lands with even more force than usual. You can't borrow virtue from a neighbouring version of yourself. This version of you is the one who's autonomous, and the one who's answerable.

You can't say "some other version of me got it right" and use that to excuse harm in this one. You can't hand the burden to a different timeline either. Sartre captured that particular flavour of responsibility well. Writing in *Existentialism is a Humanism*, in Philip Mairet's 1948 translation, he put it about as plainly as it can be put: man is nothing else but what he makes of himself. Existence comes first, and then you build whatever follows out of what you actually do, not out of what some other version of you happens to be doing somewhere else. Your decisions, large and small, are what define you. The Multi-Verse may hand you endless mirrors. Only this one shows what you actually did.

Sometimes we picture a better self, richer, fitter, more admired. But if time really is block-like in structure, and if every somewhen you've lived endures the way Chapter 2 suggested, then the choices you make now are the ones you're stuck living with. Not once. Permanently.

We've spent this whole book on a species built for short-term survival. The instincts that kept us alive, fight, hoard, defend, seduce, still run quietly in the background of everything we do. But we aren't bees. We aren't locked into a single hive-mind script. We can override instinct. We can question the wisdom of the crowd. And we can ask a harder question than either instinct or crowd wisdom usually bothers with: what kind of person do I actually want to be, here, now, in this one life I'm aware of living?

We're taught to compare ourselves constantly. Look down and you feel pity, or guilt. Look up and you feel envy, inspiration, sometimes despair. Culture monetises the whole cycle, whispering that the best self is the most admired one, the most productive one, the most envied one. If we accept the Omni-Verse, if we accept that a great many versions of us exist across the full range of possibility, comparison takes on a strange new weight. It can go one of two ways. It can cheapen this life, since some other version of you will presumably handle it. Or it can sanctify this life, since only this version of you can do what only you can do, here.

In an infinite spread of possibility you can always find a version of yourself to envy, and always find one to pity. Neither is the point. What matters isn't winning some imaginary league table of selves. It's living a life this particular you would be willing to live again.

Nietzsche posed something close to this as a thought experiment in *The Gay Science*, published in 1882. He imagined a demon slipping into your loneliest moment to tell you that this exact life, down to every detail, every joy and every unbearable pain, would return to you again and again, forever, in exactly the same sequence. His question wasn't really about physics. It was a test of how you'd receive the news. Would it crush you, or would you fall to your knees grateful for it? The Omni-Verse doesn't need to settle whether time actually loops to make that test useful. You can't control every thread running out from this moment. You can still choose to live this one as though it mattered enough to survive the demon's visit.

So who do you actually want to be, in this version of your life?

Not the version who'd win the most admiration, or collect the most trophies or followers. The version you'd most want to be you. Picture the kindest, most balanced, most fully realised version of yourself you can manage. How far off is that person from the one reading this sentence right now? And if the gap is wide, what's one small step that would close a bit of it?

Society encourages us to chase goals without asking what kind of person we become in the chasing. We dream of wealth and quietly leave out the cost, the stress-induced ulcers, the missed birthdays, the isolation, the shallow connections that money alone can buy. We might just as easily picture the selfless activist chained to a tree and skip over their struggle to pay rent, or provide stability for their children, or survive the burnout that comes from fighting the same battle every day. Chasing the shadow of imagined success risks leaving you hollow long before you catch it. And still we chase, because the more balanced alternative can feel like failure by the scoreboard everyone else is using.

Kierkegaard warned about exactly this trap in *The Sickness Unto Death*, in the translation by Howard and Edna Hong. The greatest hazard of all, he wrote, is losing one's self. What makes it dangerous is how quietly it can happen, with none of the fuss that would accompany losing an arm, a wife, or five dollars. For Kierkegaard, real despair wasn't suffering. It was living as someone you aren't. The Multi-Verse hands you an infinite number of false selves to envy from a safe distance. Becoming yourself, quietly and without pretence, is the harder and the only worthwhile task.

The Omni-Verse doesn't exist to excuse bad choices, and it doesn't exist to burden you with some imperative to be any one particular thing. It simply is. Within it, this version of you is a fully conscious, essential being. If you want your path to look different, you can reflect on that. You can pivot. You can choose differently starting now.

You don't have to be the most famous, or the most admired, or the wealthiest version of yourself. You don't need to win a race that isn't actually the one you care about. You only need to ask yourself, plainly: if this life looped forever, if this were the version I lived again and again, would I be proud to live it again? Are there parts that would fill me with shame? And if so, are they occasional, or are they daily?

A small practice: The Loop Test.

I like to sit with five questions for five quiet minutes, from time to time.

If this week repeated forever, would I be proud of how I behaved?

How would a better version of me have behaved instead?

What could I change tomorrow to take one step toward that better path?

What am I avoiding that a better version of me would face head-on?

Have I balanced my time and attention this month in a way that shows my loved ones how I actually feel?

The gift of the Omni-Verse isn't perfection. It's possibility. For every reality where you keep acting exactly as you have so far, there are realities where you reflect and change course. Some make small adjustments. Some transform entirely. You aren't a draft version of some final you. You're walking a path with more than one next step available, and each choice you make further defines who you're becoming.

One thing worth holding onto clearly. In an infinite Omni-Verse, every possible version of you exists. But "possible" is doing serious work in that sentence, and it doesn't mean anything goes. Within our Macro-Verse, the family of universes sharing our particular rulebook, gravity attracts, light travels at exactly 299,792,458 metres per second, and entropy doesn't spontaneously decrease. The Multi-Verse, our fifth-dimensional branching, varies outcomes while keeping that rulebook fixed. Within those branches there's no world where a wizard actually flies on a broomstick by magic that overrides physics itself. A rule system like that would break the macro-rules you and I actually live under, which puts it on a different branch of the ladder entirely, if it belongs on the ladder at all.

I'd encourage you to imagine the different paths open to you, the different people you could become in the next chapter of your own life. Don't fixate only on the surface changes. Weigh the good against the cost in each one. What are you actually trading off? Which of these versions could you manifest through changes you're capable of making starting today?

All of this assumes I think of myself as a one-reality kind of person, which is worth examining a little further. Biologically, most of the cells in your body get replaced over time. Skin regenerates every few weeks. Red blood cells turn over within months. Many tissues renew completely within seven to ten years. The child-body you remember is long gone. You persisted anyway. What continues isn't the material. It's a pattern: memory, temperament, story, the continuity of experience threading through time. You could argue we're little more than clones of our past selves, sharing some memories and calling it identity. Almost nobody actually thinks of it that way. We treat our past and present selves as simply, continuously, us.

Why not stretch that same reasoning a little further? Perhaps the you in another timeline, the one who made a different choice at the fork, isn't a stranger. Perhaps it's another instance of the same broader pattern, a version of you unfolding along a different branch. To function inside any one timeline, we have to perceive only that timeline while we're acting in it. Maybe the same constraint holds for time generally. A being aware of every one of its moments at once might be tempted to keep altering them, and in doing so, undo the very self that was trying to act in the first place. Nature seems to have built in a useful limit: awareness local enough to actually live, choose and mean something, right where and when you happen to be.

None of this is an argument for detachment. It's closer to the opposite. The choices you make here matter. They shape the version of reality you actually live in, and you're accountable to this timeline, these people, these consequences. You can't talk your way out of a parking ticket by claiming to be a clone of your past self, or by insisting you paid the meter in some other branch. And yet, when my own life draws to a close, I hope I can look back and recognise that nothing in the larger scheme was ever truly undone. No love was truly lost. No word went permanently unsaid. Somewhere and somewhen, across the full expanse of time and possibility, every moment endures.

I don't need to weigh one life against anyone else's and decide whether it measured up. I don't need to earn my immortality through the approval of gods, or through a legacy loud enough to outlast me. I already have it, by existing, now and permanently, as part of an infinite Omni-Verse of Somewhens.

The baby who cries when a toy disappears eventually learns object permanence, and calms. Adults can learn Temporal Permanence, and with time, learn to grieve without slipping into nihilism alongside it. The last time you laughed with someone you love is a page that stays written. So is the first time you met them. Nothing in this model erases sorrow. It reframes it. We mourn the end of shared overlap on this particular branch, and we give thanks for the pages we get to keep, permanently.

With all that in mind, I hope to meet my own end the way I met my beginning, as one more moment inside an infinite multitude of moments.

Like a child called in from play at dusk, I expect I'll feel a pang of resistance when my own time comes. That's only proof I enjoyed the day. I lived. I loved. I was loved.

I know I'll be missed by the people I connected with, and I know my absence, my temporal absence, will cause real pain. I hope they can also hold onto the fact that I'm not gone from existence altogether. We shared a path. Mine simply doesn't extend into the next part of theirs, not in this particular reality.

What actually matters is that in this reality, we got to share time and make a connection that, I hope, made both our lives a little better. I got to have them in mine. The properly good news, if this picture is even close to right, is that we'll likely get to do it again, a great many more times. An infinite number of firsts. An infinite number of lasts. All the magic that happens in between.

If time loops, or if it doesn't, you still meet yourself in the mirror of your own choices. Let that mirror show you someone you recognise. Curious. Honest. Trying.

You don't need to be the best possible version of yourself across the whole Omni-Verse. You only need to be the truest possible version of yourself in this room, on this day, in this particular chapter of your life. If that means calling your sister, or apologising, or taking a twenty-minute walk, or showing up to vote, or reading to a child, or finally booking the appointment you've been putting off, then do the small, definite thing.

All of it, everything we shared, remains. That, in the end, is what gives this one life, this single thread, its infinite weight.

12. HUMANITY IN THE OMNI-VERSE

Chapter 11 offered a kind of personal comfort, an invitation to find peace in permanence and possibility. This chapter is where we stop and weigh what that comfort actually costs. An infinite Multi-Verse can console you on a personal level. It can also be misused to excuse harm on a societal one. Like most powerful ideas, it carries a shadow, and the shadow deserves as much attention as the light.

The first time you really sit with infinite timelines, endless chances, endless versions of yourself, the temptation is obvious. Detach. If every path unfolds somewhere, does it matter what you choose in this one? If you fail someone here, isn't there a branch where you made it right? If you pollute, or destroy, or oppress in this world, won't justice prevail somewhere else regardless?

That's one of the most dangerous mistakes available to us.

It's worth being honest about why the mistake is so tempting in the first place. Every idea powerful enough to comfort you is also powerful enough to be misused, and this one is more powerful than most. A belief system that can survive contact with grief can just as easily survive contact with cowardice, if you let it. The Multi-Verse doesn't come with a built-in defence against being used that way. We have to supply the defence ourselves, and that starts with simply naming the failure mode before it has the chance to feel reasonable.

An infinite number of outcomes existing doesn't make this one less real. If anything it makes it more sacred, and more fragile. This timeline, this particular node of existence, isn't a draft. It isn't a rehearsal. It's one of infinitely many precious manifestations of what could be, and what you do here echoes, not because it alters the other branches but because it defines this one.

You can't contest a parking ticket by pointing at a version of you who paid it. In the same way, you can't excuse cruelty, neglect or destruction here on the grounds that other universes turned out better. This is the branch you're accountable for. This is our mess, and we're the ones who need to clean it up.

There's a second error waiting right behind the first, and it's subtler. It's the assumption that our survival to date guarantees our survival tomorrow. Hierarchical cultures, religious or secular, slide easily into a feeling of having been chosen. We descend from the people who survived wars and plagues, so we quietly underestimate how dangerous those things actually were. We assume the system works because we're here to watch it working.

Our inability to see the whole Omni-Verse of possible alternative outcomes creates exactly the bias you'd expect. Good fortune to date starts to feel like a promise of good fortune tomorrow, and it isn't one.

In an Omni-Verse, that logic collapses completely. For every reality where humanity thrives, there's another where it's already gone. There are histories where Hitler succeeded. Histories where the Black Death never receded. Futures where climate action arrived too late, where fascism resurged unchecked, where a pandemic outpaced public health entirely, where scientific insight simply drowned beneath misinformation.

Far from protecting us from calamity, an infinite spread of futures guarantees that the worst outcomes arise somewhere, exactly as reliably as the best ones do. This could easily be one of the doomed timelines, if we don't act like it matters that it isn't.

The comedian Jimmy Carr has a routine built on exactly this blind spot. After the COVID-19 pandemic, he asked an audience whether the precautions had felt excessive, and they cheered in agreement. His follow-up landed the joke by pointing out, in his own dry way, that of course the people still around to complain about the precautions were the ones the precautions had worked for, or who'd simply been lucky. Whichever it was, they weren't exactly a neutral panel.

It's a sharp routine and it's usually filed under confirmation bias, but that's not quite the mechanism doing the work. Confirmation bias is looking for evidence that supports what you already believe. What's happening in that room is something narrower and stranger. The only people available to have an opinion on whether the restrictions were too much were, by definition, the people who'd survived to have one. Anyone the virus took never got a vote. That's **survivorship bias**, and it's worth keeping the two separate, because survivorship bias is quieter and considerably harder to notice from inside it.

Their survival doesn't prove the restrictions were excessive. It proves they were either lucky or protected. Across the wider spread of the Multi-Verse, plenty of people in that exact room didn't survive in every branch. Plenty of versions of Jimmy Carr didn't either. In most of them, if you want the fuller joke, he was probably an accountant. If we carried the knowledge of our less fortunate counterparts around with us the way we carry our own survival, we'd feel very differently about our odds going forward. That's precisely why it's called a bias and not a fact.

There's a term I've been building toward without properly handing it to you, and it belongs here more than anywhere else in the book.

Call it **Moral Gravity**. It's the ethical weight that comes attached to knowing infinite realities exist.

The temptation is to let that knowledge pull you toward apathy, since so much is happening elsewhere regardless of what you do. Moral Gravity is the opposite pull. It means that when you're faced with a choice, the choice still matters, and arguably matters more instead of less. It's a reminder to be a careful steward of the one reality actually entrusted to you. Not because it's the only one that exists, but because it's the one that's yours to look after. We've been entrusted to each other, on this particular branch, and nowhere else carries that particular obligation for you.

Think of it the way you'd think about a single afternoon with an elderly parent. Knowing there might be other versions of that afternoon playing out in other branches doesn't lighten this one. It's still the only afternoon you get with them, in the only life you'll ever actually live through. Moral Gravity is what happens when you let that fact carry its proper weight instead of shrugging it off onto some other version of you, somewhere else, who isn't the one holding their hand.

You can't outsource your ethics to the Omni-Verse. You can't delegate care to another timeline, and you can't assume you belong to some specially protected, chosen branch. If there's a divine hand steering any of this, it's working through every path at once, the ones that flower and the ones that fail alike. We aren't chosen. We're simply present, and presence carries its own duty.

Is it all doom, then? No. On the wider scale, there are branches where we take up artificial intelligence responsibly, for human and ecological benefit instead of against it. Branches where polarisation eases, where rights expand, where resources get stewarded with something like care. There are also branches where none of that happens. The Multi-Verse doesn't promise a happy ending anywhere in particular. It promises that every possible outcome happens somewhere within it. What tilts the odds of which one we personally land in is us.

Given that survival is fairly binary, you either make it through or you don't, there may well be roughly as many ways to thrive as there are to fail. The difference sits entirely in what we decide and build together on this branch. Every decision counts toward that, including the small ones.

If we could actually see the neighbouring Multi-Verses around us, the ones so similar to ours except that humanity had already died out in them, we might not feel quite so safe here. If you could see the countless versions of yourself already gone, taken by war, famine, cosmic accident or disease, your confidence might look rather different.

Picture a forest burnt to the ground, with a single tree still standing. Perched in it, a talking squirrel proclaims that fire is nothing to worry about, since his tree has made it through so far. He might even announce that his tree was chosen. It's a ludicrous image, and it's exactly the shape of the false confidence we draw from humanity's survival to date. Of course the version of us still standing here thinks things have gone fine. That's the only version capable of having the thought.

Think of a genuine near miss you've had, a car that stopped just short of you, or didn't quite clip you crossing the road. Now consider how many branches exist where it didn't stop in time. You were hit, and that branch of you didn't continue. Does knowing that change how confident you feel walking through traffic? Put it plainly: if this were a universe where we'd already been wiped out, none of us would be here to notice it. That doesn't make past threats any less real. It only means we were lucky on this particular branch.

So where does a higher power fit into any of this, if at all? If there's a divine hand steering reality through the Omni-Verse, the most consistent reading is that it's ensuring every possible outcome arises somewhere, so that every branch of reality gets to exist. Our survival so far doesn't mean we were chosen for it. We aren't destined for reward, and we aren't exempt from consequences as we move forward from here. Every outcome that could still happen from this point will happen, to some version of us. For all we actually know, we're surrounded by a charcoal plain of branches where things went the other way.

Let's scale the same logic up to the size of a galaxy, and take a look at one of science's genuine mysteries: the Fermi Paradox.

Sometimes called the Great Silence, the Fermi Paradox points out that the numbers say the Milky Way should be teeming with life. Hundreds of billions of stars, a large share of them with planets, plenty in temperate zones. Civilisations could arise millions of years apart from one another and still have time to spread, or at least to signal. And yet, as far as we can tell, silence. No unambiguous signals, no obvious megastructures, no self-replicating probes glinting anywhere in the starlight.

The physicist Enrico Fermi is remembered, in an account colleagues gave years afterward instead of a line anyone wrote down at the time, for cutting through a lunchtime conversation about flying saucers with a single question: "Where is everybody?" The precise wording is disputed, but the question itself has outlived the argument about its phrasing.

One popular attempt at an answer proposes a **Great Filter**, some step or event that wipes out most life before it advances too far. War, climate collapse, engineered pandemics, misaligned artificial intelligence, any of these could plausibly end an advanced species before it ever goes interstellar. Others argue the filter is behind us already, and that we got lucky clearing it. Some suggest alien civilisations stay quiet on purpose, in what's sometimes called a **dark forest**, where every intelligent species keeps silent to avoid drawing the attention of something worse. Others suggest we simply evolved early, ahead of most of the pack. And some suggest intelligent life is just rarer than we assume, and that Earth's good fortune, and ours, is scarcer than it feels from inside it.

None of these candidates cancel each other out, which is part of what makes the puzzle so stubborn. The filter could be behind us and still be ahead of us again, in a different form, for the same species. Getting past one bottleneck doesn't retire the concept. It just moves the question further down the line, to whichever bottleneck comes next.

That last idea sits unusually well beside the Many-Worlds picture from Chapter 6. In a branching reality, observers necessarily find themselves in the branches where the preconditions for observation actually held. If humanity was wiped out early in ninety-nine branches out of a hundred, you and I would still wake up in the one remaining branch, because that's the only branch where anyone's left to ask the question at all. Our existence is evidence that survival was possible. It isn't evidence that it was likely. Seen that way, our own presence quietly inflates how common we assume long-lived, spacefaring civilisations must be. The squirrel on the last standing tree makes for a poor fire statistician. Maybe the silence out there means the odds of any given tree burning down were considerably worse than we'd like to think. Maybe our galactic neighbours simply weren't as lucky as we've been.

So how do we actually work against a bias this deeply built in?

Start by recognising it exists at all, since that's most of the battle. It tends to be the most comfortable among us who feel chosen most easily, which means the rest of us carry a particular responsibility to be the counter-voice. We aren't chosen. We're accountable. Taking our place in the Omni-Verse seriously means acting with real duty and real care, and steering toward the future we actually want instead of assuming it will simply arrive on its own.

If a meteor turns up tomorrow, one big enough to end things outright, we can take genuine comfort in knowing our final days here are exactly as enduring as our first ever were. Short of an ending like that, one we hold no blame for, we carry an ordinary duty of care instead: to learn plainly from our past as we shape whatever comes next, together.

The fewer branches that make it through, the more sacred every surviving branch becomes.

Stewardship was never a luxury reserved for a multiversal view of things. It's a local duty, due right here, on this particular branch, starting now. In the Omni-Verse it was never a case of anything goes. It's a case of every decision counting, because it's the only one you'll ever actually get to make.

None of this asks you to carry the weight of every branch at once. That would be an impossible burden, and it isn't the one being offered here. You carry the weight of one branch, the one you're actually standing on, and that's already a full load for any single life. The infinite versions of you elsewhere are handling their own branches, or failing to, on their own terms. Yours is the only ledger you'll ever actually be asked to balance, in this life, on this branch, with these people.

Perhaps that's the quietest gift buried in all of this. Not permission to relax because the outcome is guaranteed somewhere regardless of what you do, but a clear boundary around exactly what's yours to do something about. Everything inside that boundary counts completely. Nothing outside it was ever yours to carry in the first place.

FINAL REFLECTION

We've covered a great deal of ground since that first chapter about building an idea-system. Branches splitting endlessly outward. Universes folding back on themselves and mirroring their own beginnings. Laws that vary from one tree to the next, and a forest holding all of them. Ancient cosmologies reaching for the same handful of metaphors our own equations reach for now. None of it was meant to add up to a single verdict. It was meant to give you enough material to build something of your own out of it.

I was clear at the outset that this book isn't a creed, and it isn't an ideology. It's an exploration of individual ideas, taken alone and then in combination, to see what they might look like inside an **Idea-System** of your own. I kept reading while I wrote it, and I updated chapters when new research supported or challenged what I'd already drafted. My honest sense is that a broad and still-evolving consensus supports a good number of the ideas you've just read. That seems reason enough to understand them, and to consider whether they've earned a place in your own thinking. More than that, though, I hope you'll hold onto the right to question popular thinking in whatever guise it turns up in.

I made it clear early on that I don't take issue with any specific religion, and I want to close on the same note. I identify as an atheist myself, without a religion of my own, but I don't question anyone else's right to hold religious belief, provided it isn't being imposed on someone who doesn't share it. Religious beliefs, like scientific ideas, are still ideas. Some scientific theories look solid for a while and then don't survive scrutiny. The same willingness to keep questioning has to apply everywhere, including to the things we hold most dear. You have to be willing to let go of the wooden raft if you actually want to swim.

I'm not a spiritual guru or advisor, and I wouldn't claim to be one. I do recognise the real strength and comfort a lot of people draw from spiritual practice, and I don't think that comfort is worth dismissing. With that in mind, I want to say plainly that nothing in this book closes the door on God, or gods, if you want to include one in your own idea-system. Do I personally picture that as an old man in a toga, or a figure rising from a shell? I don't, but that's beside the point. What I picture isn't the question that matters here. Whatever shape the idea takes for you, the physics in these chapters was never built to argue you out of it. Most of it leaves the question exactly where it found it.

Could the Omni-Verse itself be an expression of some larger consciousness? Or is consciousness the thing doing the exploring, working its way through every possible outcome the Omni-Verse contains? Quite possibly. Not necessarily. But possibly, and I'd rather leave that door open than pretend I know which side of it is true.

The point I keep circling back to is this. Rigid ideologies, religious or scientific, can offer real comfort, and so can junk food. Neither one is doing you any favours in the quantities that feel best in the moment. An overly rigid belief tends to crowd out your own objective reasoning. Treat any doctrine as absolute and you stop questioning it. Hand your own judgement, your own curiosity, your own sense of right and wrong over to someone else entirely, and you've handed them a fair amount of control over you in the process. You've also given away one of the better parts of being human.

Flexibility isn't the same as having no convictions at all. An idea-system can still hold firm views. The difference is that it holds them the way a good scientist holds a theory, with real confidence right up until better evidence turns up, and with the door left open rather than bricked shut. That's a harder way to live than simply picking a side and staying there. I'd still argue it's the more honest one.

If we turn away from the big questions because they seem to contradict what we already believe, or because they feel too large to properly imagine, we rob ourselves of something worth keeping. The right to wonder at the sheer improbability of any of this existing at all.

Perhaps you agree with everything in these chapters. Perhaps you don't, and you've spent the whole book arguing back at me in your head. Either way is fine. That was rather the point of building an idea-system instead of handing you a belief system to accept whole.

None of this needs to live only in the abstract. The Loop Test from Chapter 11 and the weight I called Moral Gravity in Chapter 12 were both attempts to bring the ideas back down to something you could actually use on an ordinary Tuesday. You don't need to resolve whether time loops or branches to ask yourself, tonight, whether you'd be glad to live this week again exactly as you lived it. That question works whichever model turns out to be closer to true.

I'll leave you with one question instead of a conclusion. At the end of this life, looking back over everything, which path would be the one you'd most want to walk again?

I hope you enjoyed the swim.

GLOSSARY

Part A: Terms and Concepts

Alt-Multi-Verse. A term coined in this book for a collection of universes governed by different foundational laws of physics, in contrast with the Multi-Verse, where physical laws stay fixed and only outcomes vary. Introduced in Chapter 7 and developed further in Chapter 9.

Anthropic Principle. The observation that we find ourselves in a universe capable of supporting observers, since no other kind of universe could have anyone in it to notice otherwise. A well-established idea in physics and philosophy, not original to this book. Introduced in Chapter 7.

Anthropomorphism. The projection of human traits, motives or intentions onto non-human phenomena. Discussed in Chapter 10.

Belief System. A structured framework for interpreting reality, often resistant to revision. Contrasted with an Idea-System throughout the book, particularly in Chapters 1 and 10.

Big Bang. The standard model for the origin of our universe, describing an expansion from an extremely hot, dense early state. Discussed in Chapter 3 and given a fuller treatment in Chapter 8, where it is reframed as the end of a preceding period of cosmic inflation rather than an absolute beginning.

Big Crunch. A hypothetical scenario in which cosmic expansion reverses and the universe collapses back into a hot, dense state. Chapter 3 discusses a 2022 calculation suggesting this could begin surprisingly soon on a cosmic timescale, conditional on dark energy behaving as a dynamic quintessence field rather than a fixed cosmological constant. Linked to cyclic models in Chapters 4, 6 and 8.

Big Rip. A hypothetical end state in which dark energy strengthens over time until it overwhelms every other force, tearing apart galaxies, then stars, then atoms. Placed at roughly twenty-two billion years from now under one particular model, in Chapters 3 and 5.

Block Theory (Eternalism). A model of time in which past, present and future all exist, with time treated as a dimension rather than a process. Introduced in Chapter 2 and developed in Chapters 3, 4 and 11.

Brane Cosmology. A model in which our universe is a membrane, or brane, within a higher-dimensional space, with other branes potentially nearby. Introduced in Chapter 6 and linked to Norse cosmology in Chapter 10.

Complementarity. Niels Bohr's 1927 principle that certain descriptions of nature, such as light behaving as a wave or a particle, are mutually necessary and mutually exclusive: both are needed, and no single experiment shows both at once. Introduced in Chapter 1.

Confirmation Bias. The tendency to interpret evidence in a way that confirms pre-existing beliefs. Referenced in Chapters 1 and 10, and distinguished from Survivorship Bias in Chapter 12.

Conformal Cyclic Cosmology. Roger Penrose's proposal that the universe passes through an infinite sequence of cycles, or aeons, with the remote future of one aeon resembling the Big Bang of the next. Discussed in Chapter 3.

Cosmic Inflation. A theory describing an extremely brief period of exponential expansion in the very early universe, which smooths and flattens space and seeds the fluctuations that later became galaxies. Foundational to Chapter 8.

Cosmic Microwave Background (CMB). The faint thermal radiation left over from roughly 380,000 years after the Big Bang, first detected by Arno Penzias and Robert Wilson in 1964. Used as evidence for inflation and early cosmic conditions in Chapters 3 and 8.

CPT-Symmetric Universe. A model in which a universe running backward in time is treated as the mirror image of one running forward, related by the combined operations of Charge, Parity and Time reversal. Introduced in Chapter 4.

Dao. In Daoist cosmology, the unseen force or Way behind all things, expressed through the interplay of yin and yang. Discussed in Chapter 10.

Dark Energy. The name given to whatever is driving the observed acceleration of cosmic expansion. Introduced in Chapter 3 and returned to in Chapters 5 and 8, where its constancy is treated as an open question.

Dark Forest. A proposed solution to the Fermi Paradox, in which intelligent civilisations stay deliberately silent to avoid drawing the attention of something more dangerous. Discussed in Chapter 12.

Decoherence. The loss of quantum interference that occurs when a quantum system interacts with its surroundings, which is what prevents branches in the Many-Worlds Interpretation from communicating with one another. Introduced in Chapter 6.

Determinism. The view that all events are causally fixed by what preceded them. Used in Chapter 11 to frame the question of free will.

Digital Physics. The proposal that the universe behaves as a computational or information-based system at its most fundamental level. Discussed in Chapter 9.

Dogma. A rigid, unquestioned belief system. Discussed throughout, particularly in Chapter 10.

Eighth Dimension. In this book's ladder of possibility, the level at which the meta-laws themselves vary, meaning the rules about what a rule is allowed to be. Introduced in Chapter 7.

Einstein-Rosen Bridge. A theoretical tunnel connecting distant points in spacetime, proposed by Albert Einstein and Nathan Rosen in 1935, more commonly known as a wormhole. Discussed in Chapter 6 in relation to black-hole cosmology.

Entropic Tension Force (ETF). A speculative force proposed by this book, describing how entropy shared across paired universes might exert a balancing or shaping influence on cosmic structure. Introduced in Chapters 4 and 5, and extended speculatively in Chapter 8.

Entropy. A measure of disorder within a system. Used across the book in both scientific and metaphorical registers, representing thermodynamic evolution, cosmic balance and, in Chapter 12, moral weight.

Eternal Return. Friedrich Nietzsche's thought experiment, from *The Gay Science* (1882), imagining that this exact life would recur infinitely in precisely the same sequence, used as a test of how fully a person is willing to affirm their own life. Applied in Chapter 11.

Existential Detox. An informal term coined in this book for the process of loosening a rigid belief in favour of open inquiry, without necessarily discarding the belief altogether. Introduced in Chapter 1.

Existentialism. A philosophy emphasising personal freedom, authenticity and moral responsibility. Referenced in Chapter 11 through Jean-Paul Sartre.

Fifth Dimension. In this book's usage, the level at which parallel universes share identical physical laws but diverge in outcome. Introduced in Chapter 6 and returned to in Chapters 7 and 11.

Fine-Tuning Problem. The observation that many independent physical constants appear to sit within a narrow range required for stars, chemistry and life to exist at all. Introduced in Chapter 7.

General Relativity. Albert Einstein's theory describing gravity as the curvature of spacetime caused by mass and energy. Referenced in Chapters 6 and 8.

Great Filter. A proposed explanation for the Fermi Paradox, suggesting some step or event reliably prevents most life from advancing to an interstellar stage, whether that filter lies behind us or ahead. Discussed in Chapter 12.

Growing Block Theory. A model of time in which the past and present exist, but the future does not yet exist. Introduced in Chapter 2.

Hawking Radiation. The theorised faint emission of radiation from black holes, predicted by Stephen Hawking, arising from quantum effects near the event horizon and causing black holes to lose mass and eventually evaporate. Referenced in Chapter 5.

Idea-System. The flexible mental model proposed by this book for holding, exploring and revising ideas without demanding permanent commitment to any one of them. Contrasted with a Belief System throughout, and introduced formally in Chapter 1.

Idealism. The philosophical view that reality is fundamentally mental rather than material. Discussed in Chapter 9 through George Berkeley and Donald Hoffman.

Kantian Ethics. Immanuel Kant's deontological system, in which moral worth lies in the will behind an act rather than its outcome, tested by whether the underlying principle could be willed as a universal rule. Discussed in Chapter 11.

Logos. In Stoic cosmology, the rational ordering force structuring the universe. Discussed in Chapter 10.

Loop Quantum Cosmology (LQC). An alternative to singularity-based cosmology, in which gravity becomes repulsive at extreme densities, replacing collapse with a bounce into renewed expansion. Discussed in Chapters 4 and 8.

Loop Test, The. A short reflective practice coined in this book, asking whether a person would be proud to live their current choices on repeat. Introduced in Chapter 11.

Ma'at. In Egyptian cosmology, the principle of truth, balance and cosmic order, weighed against the heart of the dead. Discussed in Chapter 10.

Macro-Verse. A subset of the Omni-Verse comprising all universes that share our particular physical laws. Introduced in the Introduction and returned to in Chapter 11.

Many-Worlds Interpretation. Hugh Everett's 1957 proposal that every possible outcome of a quantum measurement occurs, each in its own branch, with no collapse of the underlying possibilities. An interpretation of quantum mechanics rather than a rival to it, since it makes identical predictions. Foundational to Chapter 6.

Mathematical Universe Hypothesis. Max Tegmark's proposal that every mathematically self-consistent structure exists physically. Discussed in Chapter 9.

Maya. In Hindu and Buddhist thought, the idea that the observable world is a kind of perceptual shell over a less tangible underlying reality. Discussed in Chapter 10.

Meta-Laws. Hypothetical rules governing how universes, or the laws that govern universes, arise in the first place. Explored in Chapter 9.

Modal Realism. David Lewis's claim that all logically possible worlds are equally, concretely real, not merely possible in the abstract. Explored in Chapter 9.

Moral Agency. The capacity to make choices carrying ethical weight. Central to Chapter 12.

Moral Gravity. A term coined in this book for the ethical weight attached to knowing that infinite realities exist. It names the pull toward apathy that this knowledge invites, and the counter-pull of remaining a careful steward of the one reality actually entrusted to you. Introduced properly in Chapter 12.

Multi-Verse. As used in this book, an ensemble of parallel universes sharing identical physical laws but differing in outcome. Introduced in Chapter 6 and referenced throughout the remainder of the book.

Ninth Dimension. In this book's ladder of possibility, the level at which the whole collection of Alt-Multi-Verses is considered together, sometimes described as the forest holding every tree. Introduced in Chapter 7.

Object Permanence. The developmental understanding, typically acquired in infancy, that objects continue to exist even when they cannot be perceived. Used metaphorically in Chapter 11 to introduce Temporal Permanence.

Omni-Verse. The totality of all realities in this book's framework, encompassing every universe, timeline and dimensional framework, including those governed by different physical laws. Introduced in the Introduction and referenced throughout.

Ouroboros. An ancient symbol of a serpent consuming its own tail, representing cyclical renewal. Discussed in Chapter 10.

Panpsychism. The view that consciousness is a fundamental and widespread feature of reality, present in some minimal form throughout it. Discussed in Chapter 9 through William James, Alfred North Whitehead, Bertrand Russell and Philip Goff.

Presentism. The view that only the present moment is real. Contrasted with Eternalism and Growing Block Theory in Chapter 2.

Psychological Arrow of Time. Our subjective experience of time moving forward, with memory behind us and uncertainty ahead. Distinguished from the Thermodynamic Arrow of Time in Chapter 3.

Quantum Fluctuation. A brief, random variation in a quantum field, occurring even in what would otherwise be considered empty space. Discussed in Chapter 3 and given a fuller causal account in Chapter 8, where inflation stretches such fluctuations to cosmic scale, seeding every structure in the universe.

Quantum Immortality. A thought experiment, sometimes called quantum suicide, holding that because a person can only ever observe branches in which they survive, their experience would always continue somewhere. Distinct from the broader argument this book makes about permanence, and addressed briefly in Chapter 6.

Reverse-Verse / Counter-Verse. Speculative universes, developed across Chapters 3, 4 and 6, in which time runs backward relative to our own, or which originate from the same shared boundary as ours. Central to the development of the Entropic Tension Force.

Sacred Dualities. Paired, interdependent opposing forces, such as yin and yang or creation and destruction, appearing across ancient cosmologies. Discussed in Chapter 10.

Scalar Field / Inflaton Field. The hypothesised field whose potential energy is thought to have driven cosmic inflation, decaying into matter and radiation at the field's end. Discussed in Chapter 8.

Seventh Dimension. In this book's ladder of possibility, the level at which both the rulebook and the starting template of a universe vary, including its boundary conditions and large-scale geometry. Introduced in Chapter 7.

Simulation Theory. The hypothesis that our reality might be a simulation run by an advanced intelligence. Nick Bostrom's 2003 formulation is a trilemma rather than a flat claim: that civilisations reliably go extinct before developing the capacity to run such simulations, or reliably choose not to, or we are almost certainly living in one. Discussed in Chapter 9.

Sixth Dimension. In this book's ladder of possibility, the level at which different universes share an overarching physical framework but vary in their specific rules or constants. Introduced in Chapter 7.

Somewhen. A term coined in this book for other genuinely existing moments in time, applying the logic of object permanence to time itself. Introduced in Chapter 2 and central to Chapter 11.

Straight Universe. A term coined in this book to describe a spatially flat universe with Euclidean, non-curving large-scale geometry, chosen in place of the more common but misleading term "flat universe." Introduced in Chapter 8.

String Theory Landscape Problem. The difficulty of selecting among an extremely large number of possible string theory configurations, commonly estimated around 10^500, each implying a different set of physical laws. Discussed in Chapter 9.

Survivorship Bias. A cognitive error in which only the surviving examples of an event remain available to be considered, skewing judgement of that event's danger or rarity. Distinguished from Confirmation Bias in Chapter 12.

Temporal Permanence. A term coined in this book for the idea that all points in time genuinely exist, whether or not they are currently being perceived, extending the logic of object permanence to time. Central to Chapter 11.

Thermodynamic Arrow of Time. The direction in which entropy increases, used to define one basis for the forward flow of time. Discussed in Chapters 3 and 8.

Time Dilation. The relativistic effect by which time passes at different rates depending on relative gravity or velocity. Referenced in Chapter 6.

Wormhole Theory. The proposal that tunnels through spacetime, also known as Einstein-Rosen bridges, might connect distant regions of a universe or separate universes entirely. Referenced in Chapter 6.

Part B: People Referenced

Alan Guth. Physicist; originator of the theory of cosmic inflation. Chapter 8.

Albert Einstein. Physicist; developed General Relativity and, with Nathan Rosen, proposed the Einstein-Rosen Bridge. Chapters 6 and 8.

Alfred North Whitehead. Philosopher and mathematician; associated with early panpsychist thought. Chapter 9.

Alfred Wegener. Geophysicist; proposed continental drift in 1912 and was dismissed for roughly fifty years before vindicating evidence arrived. Chapter 10.

Andrei Linde. Physicist; extended inflationary theory into eternal and chaotic inflation. Chapter 8.

Arno Penzias and Robert Wilson. Physicists; discovered the Cosmic Microwave Background in 1964. Chapter 3.

Barry Marshall. Physician; drank a culture of Helicobacter pylori in 1984 to demonstrate that bacteria, not stress, caused stomach ulcers, and shared the 2005 Nobel Prize in Physiology or Medicine with Robin Warren. Chapter 10.

Bertrand Russell. Philosopher and logician; associated with early panpsychist thought. Chapter 9.

David Lewis. Philosopher; originator of Modal Realism. Chapter 9.

Donald D. Hoffman. Cognitive scientist; argues that the physical world functions as an interface rather than as fundamental reality. Chapter 9.

Edward Fredkin. Computer scientist and physicist; early proponent of digital physics. Chapter 9.

Edwin Hubble. Astronomer; established the expansion of the universe through redshift measurements. Chapter 8.

Emmy Noether. Mathematician; proved in 1918 the theorem linking every conservation law to an underlying symmetry. Chapter 7.

Enrico Fermi. Physicist; remembered for posing the question behind what became known as the Fermi Paradox. Chapter 12.

Ethan Siegel. Science writer; writes on cosmology and the implications of inflationary theory. Chapter 8.

Friedrich Nietzsche. Philosopher; proposed the thought experiment of Eternal Return in *The Gay Science*. Chapter 11.

Galileo Galilei. Astronomer; pioneer of observational astronomy. Chapter 10.

George Berkeley. Philosopher; founder of subjective idealism. Chapter 9.

Georges Lemaître. Physicist and priest; proposed the primeval atom, an early ancestor of Big Bang cosmology, in the 1920s. Chapter 8.

Henrietta Leavitt. Astronomer; discovered the period-luminosity relationship that made reliable cosmic distance measurement possible. Chapter 8.

Howard and Edna Hong. Translators of Kierkegaard's *The Sickness Unto Death* into English. Chapter 11.

Hugh Everett. Physicist; proposed the Many-Worlds Interpretation in 1957. Chapter 6.

Hypatia. Philosopher and astronomer of late antiquity. Chapter 10.

I. J. Good. Mathematician; early contributor to black-hole cosmology. Chapter 6.

Immanuel Kant. Philosopher; foundational figure in deontological ethics. Chapter 11.

Isaac Newton. Physicist; formulated the laws of motion and universal gravitation. Chapter 10.

Jean-Paul Sartre. Existentialist philosopher; author of *Existentialism is a Humanism*. Chapter 11.

Jimmy Carr. Comedian; his COVID-era routine is discussed as an illustration of survivorship bias. Chapter 12.

Max Tegmark. Physicist; developed the Mathematical Universe Hypothesis. Chapter 9.

Nathan Rosen. Physicist; co-author, with Albert Einstein, of the Einstein-Rosen Bridge. Chapter 6.

Nick Bostrom. Philosopher; proposed the simulation argument in its trilemma form. Chapter 9.

Nikodem Popławski. Physicist; revived and extended black-hole cosmology using Einstein-Cartan gravity and torsion. Chapter 6.

Philip Goff. Philosopher; contemporary advocate of panpsychism. Chapter 9.

Philip Mairet. Translator of Sartre's *Existentialism is a Humanism* into English in 1948. Chapter 11.

Pope Gregory I. Historical figure linked to early Catholic dietary traditions. Chapter 10.

Raj Kumar Pathria. Physicist; proposed early black-hole cosmology models, sometimes called Schwarzschild cosmology. Chapter 6.

Robin Warren. Pathologist; shared the 2005 Nobel Prize in Physiology or Medicine with Barry Marshall for establishing the bacterial cause of stomach ulcers. Chapter 10.

Roger Penrose. Cosmologist; proposed Conformal Cyclic Cosmology. Chapter 3.

Rudolf Clausius. Physicist; formulated the second law of thermodynamics. Chapter 3.

Søren Kierkegaard. Philosopher; wrote on despair and authentic selfhood in *The Sickness Unto Death.* Chapter 11.

Stephen Hawking. Physicist and cosmologist; predicted the radiation that carries his name. Chapter 5.

Stephen Wolfram. Physicist and computer scientist; proponent of digital and computational models of the universe. Chapter 9.

Vesto Slipher. Astronomer; made early measurements of galaxy redshifts. Chapter 8.

William James. Philosopher and psychologist; associated with early panpsychist thought. Chapter 9.

Part C: Cited Works and Belief Systems

Back to the Future (film). Used to illustrate the logical paradoxes of conventional time travel. Chapter 6.

Chaotic Inflation. Andrei Linde's model of self-generating, potentially eternal inflation. Chapter 8.

COBE / WMAP / Planck. Successive space missions that measured the Cosmic Microwave Background with increasing precision, confirming several predictions of cosmic inflation. Chapter 8.

Conformal Cyclic Cosmology. Roger Penrose's proposal that the universe passes through an infinite sequence of cycles. Chapter 3.

Cosmic Inflation Theory. Alan Guth's 1980 theory explaining the uniformity and large-scale structure of the universe. Chapter 8.

Daoism. Chinese cosmological tradition centred on flow, harmony and the interplay of yin and yang. Chapter 10.

Egyptian Cosmology. Ancient tradition centred on Ma'at and the judgement of the soul after death. Chapter 10.

Existentialism is a Humanism. Jean-Paul Sartre's 1946 lecture, published in Philip Mairet's 1948 English translation, setting out existentialist ethics. Chapter 11.

The Gay Science. Friedrich Nietzsche's 1882 work containing the thought experiment of Eternal Return, in section 341. Chapter 11.

General Theory of Relativity. Albert Einstein's foundational theory of gravity and spacetime. Chapters 6 and 8.

Groundwork of the Metaphysics of Morals. Immanuel Kant's foundational work of moral philosophy. Chapter 11.

Hindu Cosmology. Tradition describing vast cycles of cosmic time, including the concept of Maya. Chapter 10.

Loop Quantum Cosmology. A bounce-based alternative to Big Bang singularity cosmology. Chapters 4 and 8.

Mathematical Universe Hypothesis. Max Tegmark's proposal that all mathematically consistent structures exist physically. Chapter 9.

Mayan and Aztec Cosmologies. Mesoamerican traditions treating time as cyclical and cosmically significant. Chapter 10.

Norse Mythology. Tradition describing the World Tree, Yggdrasil, and cyclical destruction and renewal through Ragnarök. Chapter 10.

The Sickness Unto Death. Søren Kierkegaard's 1849 work on despair and authentic selfhood, cited here in the Hong and Hong translation. Chapter 11.

The Simulation Argument. Nick Bostrom's 2003 trilemma concerning whether we are living in a simulation. Chapter 9.

Stoicism. Greek philosophical tradition centred on the Logos and cosmic determinism. Chapter 10.

About the Author

Patrick Wyliams writes on science and philosophy. He is also an entrepreneur, and his companies have worked in risk management, financial technology and threat evaluation. Having spent much of his career at the intersection of philanthropy, social investment and sustainability, he has contributed to a number of economic reports and has campaigned publicly for reform in social inclusion, sustainability and social justice policy.

He is married, is a father, and lives in Ireland, where he prefers to keep a private life. Patrick Wyliams is a pen name, adopted so that the work can stand on its own.

www.ingramcontent.com/pod-product-compliance
Lightning Source LLC
LaVergne TN
LVHW022036200926
848481LV00003B/707
* 9 7 9 8 2 3 1 3 6 6 1 7 0 *